The Learning Imperative

The Harvard Business Review Book Series

The Learning Imperative

Managing People for Continuous Innovation

Edited, with
an Introduction by
Robert Howard
Foreword by
Robert D. Haas

A Harvard Business Review Book

The *Harvard Business Review* articles in this collection are available as
individual reprints. Discounts apply to quantity purchases. For information
and ordering contact Operations Department, Harvard Business School
Publishing Corporation, Boston, MA 02163. Telephone: (617) 495-6192, 9
a.m. to 5 p.m. Eastern Time, Monday through Friday. Fax: (617) 495-6985,
24 hours a day.

The paper used in this publication meets the requirements of the American
National Standard for Permanence of Paper for Printed Library Materials
Z39.48-1984

Library of Congress Cataloging-in-Publication Data

The learning imperative : managing people for continuous innovation /
edited, with an introduction by Robert Howard ; foreword by Robert D.
Haas.
        p.   cm. — (Harvard business review book)
    Includes bibliographical references and index.
    ISBN 0-87584-432-4
    1. Personnel management—United States—Case studies.
    2. Technological innovations—United States—Case studies.
    I. Howard, Robert, 1954-   . II. Series: Harvard business review book
series.
    HF5549.2.U5L33 1993   93-9876
    658.3′ 124—dc20   CIP

Contents

boundaries of hierarchy, function, and geography, they must learn how to manage a new set of psychological boundaries in their work relationships with superiors, subordinates, and peers.

Surprisingly, it is often the most educated, successful, and motivated individuals who find it the most difficult to learn, using their analytical skills as a defense against acknowledging or learning from their mistakes. Real learning is possible only when professionals begin to identify the gaps between their intended and their actual behavior.

The Levi Strauss Aspirations Statement is a direct attempt to tap into the more personal dimension of work in the new economy. Levi chairman and CEO Robert Haas describes how managers are using the values articulated in the statement to set standards; shape behavior; and give more accountability, authority, and information to people on the front line of the business.

Part IV   Getting from Here to There

Most managers understand the need for organizational change but don't know how to bring it about. They assume that change must start at the top with companywide programs designed to alter formal policies and structures. In fact, the critical path to corporate renewal is to direct a nondirective change process, one that starts at the grass roots and emphasizes ad hoc solutions to real business problems.

Foreword

Robert D. Haas

These are turbulent times for business. Companies must engage in constant renewal to succeed. Relentless change demands relentless improvement.

The Learning Imperative is a survival manual for corporations trying to cope with change. As CEO of a complex international corporation with more than 30,000 employees, I've found that learning is essential for both individual and corporate growth. It is also directly linked to our financial success.

At Levi Strauss & Co. we've found that there are no obvious road maps and no clear models for success in today's business environment. Like the other companies profiled in this book, we have had to struggle to reinvent our company and ourselves. Our particular approach has been to try to anticipate change through continuous improvement in every aspect of our business. We are restructuring our sewing facilities to replace traditional assembly processes with self-managed teams. We are designing new pay programs to encourage collaboration and teamwork. And we are speeding up our product cycles and reducing manufacturing lead times to improve customer service.

Because our people are absolutely essential to this continuous-improvement process, we are also investing in their growth and skill development through values-based education programs. We want to create a workplace where employees are empowered to make a difference. We are trying to foster an organizational climate where people share information openly and broadly, listen to each other's point of view, and are able to admit it when they don't have all the answers.

We've found that our company values help create a learning culture and provide an effective framework for making decisions.

I don't want to mislead you. Building a learning organization is hard work. Incorporating continuous improvement into everything we do, encouraging employees to acquire new skills and work differently, and empowering them to make decisions never happen easily or quickly. We are engaged in a long-term cultural transformation that requires unflagging commitment and unwavering sponsorship on the part of senior management. In particular, it challenges many managers to "unlearn" the behaviors that previously made them successful.

But our experience has taught us that most people accept the challenge and enjoy the benefits of working in a learning organization. In today's world, people are better educated and better informed than at any other point in human history. As a result, employees want a greater say in the workplace. Only those managers who embrace an empowering leadership style can expect to gain the commitment of such employees and harness their energy and creativity. And the payoff of empowerment is clear: greater flexibility, responsiveness, and competitiveness.

Many management books encourage leaders to stay focused on their customers' needs and keep in touch with their own organization through "management by walking around." The articles in this book provide managers with an opportunity to walk around some of the world's leading companies and to discover how they are making the "learning imperative" an integral part of their business practice.

Introduction

Robert Howard

Every manager knows that the competitive environment is changing. Competition is getting tougher. Technological innovation is happening faster. Both customers and business partners are becoming more demanding. In so dynamic and volatile an economy, the chief source of competitive advantage is an organization's people—in particular, their ability to anticipate change, adapt to new circumstances, and invent new business practices.

Learning at all levels of the organization is not just an advantage in achieving these goals. It is an *imperative* for long-term competitive success. But this learning imperative has its own challenges for managers. They need to know how to design, build, and lead organizations that are equal to the more stringent demands of the new economy. They need to make continuous "organizational innovation" the cornerstone of their company's competitive strategy.

The Learning Imperative: Managing People for Continuous Innovation brings together for the first time 15 recent articles from the *Harvard Business Review* on the subject of organizational innovation. The anthology spells out the strategic logic, organizational design, psychological challenges, and key implementation issues of the learning organization. And it offers compelling portraits of organizational innovation as practiced at leading companies around the world, including Wal-Mart, Xerox, Levi Strauss, and Motorola in the United States; Honda, Canon, and Matsushita in Japan; and Swiss watchmaker SMH and Italian apparel manufacturer Gruppo GFT in Europe.

Although published independently over roughly a three-year period, these articles are the fruit of a common editorial agenda at the

Harvard Business Review in the early 1990s. Like many observers of the world of management, those of us on the magazine's staff at the time recognized that the economy is in the midst of a fundamental transformation. Global competition is multiplying rivals and complicating business strategy. New information technologies are reinventing products, fragmenting markets, and reorganizing work. Finally, the arrival of a new generation in the workplace is creating new demands on companies for flexibility, diversity, and more meaningful work.

We believed that these changes have profound implications for the theory and practice of management. In particular, they are making the ability to examine and think critically about *ideas* an increasingly central part of every manager's job. New competitive realities are forcing managers to question received wisdom—about strategy, corporate governance, work organization, the very purposes and practices of management—and to rethink old formulas in the light of new circumstances. Similarly, managers have to know how to test and evaluate new ideas and solutions, distinguishing what is genuinely helpful from what is mere hype. These new challenges helped define how we understood HBR's editorial mission: to be a magazine of ideas for managers, to teach them not what to think but what to think about, and in this way to be a resource for their own continuous learning.

Meeting these goals, however, required questioning some of our own received wisdom. Take a simple example: traditionally, the intellectual "division of labor" among HBR editors has corresponded more or less to the functional departments of a traditional corporation—marketing, manufacturing, strategy, information technology, and the like. But in a world where more companies are striving to break down rigid functional barriers that hinder innovation and competitiveness, this conventional way of dividing up the world, although still useful, was no longer enough. Increasingly, we found that the most provocative ideas and action were at the "interface" between the usual categories. They combined technology *and* strategy, manufacturing *and* marketing, business *and* public policy. Capturing those ideas meant redefining our traditional areas of expertise.

This was especially true in the area that became my chief responsibility when I joined the HBR staff in January 1990: human resources. On the one hand, "people" are at the very center of managerial concern in the new economy. In a competitive environment founded on constant change, one potential source of stability and continuity is an organization's employees, their distinctive ways of working together and their ability to exploit change before competitors do. And yet, the issues that this central managerial challenge raises (the impact of technology on work organization, the complexities of teamwork and organizational learning, the dynamics of change management, just to name a few) go far beyond the traditional charter of the typical human resources department. They are the responsibility of *every* manager—and especially those at the top.

So, my goal was to search for articles that conceived of organizational innovation broadly—not as some functional specialty but as the very essence of a company's business. I discovered that a number of business thinkers and senior executives at leading companies around the world were already well on their way toward spelling out the implications of precisely this point of view. In their effort to create organizations that can survive and thrive in the new economy, these thinkers and executives are redefining what it means to be a manager. *The Learning Imperative* is a record of what they have learned.

Decoding the Business Logic of Organizational Innovation

The starting point in that learning process is understanding how the new emphasis on organizational innovation is a direct response to profound changes in the competitive environment. Part I of *The Learning Imperative* includes articles that describe these changes and their implications for organizations and the people who manage them. These articles decode the "business logic" of organizational innovation. They also introduce the central themes of the entire book.

It is hard to imagine a better place to begin than with "The New Society of Organizations" by Peter Drucker. One of the most incisive observers of management in the twentieth century, Drucker addresses the key factors that will shape management in the twenty-first century. According to Drucker, the world economy is in the midst of an epochal transformation: the shift to a "knowledge society." Increasingly, knowledge is not just one new resource among many; it is *"the* primary resource for individuals and for the economy overall."

For Drucker, the knowledge society is also a "society of organizations." What distinguishes the modern organization (in business or any other sphere of life) is its single-minded focus on a common task. Organizations, writes Drucker, "put knowledge to work." They integrate a variety of specialized knowledges in order to accomplish a common organizational task. In its never-ending pursuit of its task, every organization is a "destabilizer," dedicated to "the systematic abandonment of whatever is established, customary, familiar, and comfortable" in favor of more efficient and more effective practices. In this respect, the modern organization is a platform for "social innovation" (which, Drucker reminds us, "is equally important and often more important than scientific innovation").

The essential purpose of management in the knowledge society, then, is nothing less than to encourage systematic organizational innovation. In order to do so, however, managers must first learn how to manage a fundamental tension—the tension between the organization and the people who inhabit it. Drucker makes the important point that in a knowledge economy, the true source of competitive advantage is not so much technology, R&D, or even knowledge itself. It is *people*, the knowledge workers whose skills and expertise are the foundation of all innovation. On the one hand, knowledge workers need the organization in order to put their knowledge to work. On the other hand, they (not the organization) own the chief means of production, and they can take their knowledge out the door at a moment's notice. "The more an organization becomes an organization of knowledge workers," writes Drucker, "the easier it is to leave it and move elsewhere." As a result, every organization "is always in competition for its most essential resource: qualified, knowledgeable people."

This creative tension fuels the learning imperative. According to Drucker, the only way to attract and keep the best people is to provide them with an environment that allows learning and innovation to flourish. "Loyalty can no longer be obtained by the paycheck. The organization must earn loyalty by proving to its knowledge employees that it offers them exceptional opportunities for putting their knowledge to work."

When companies discover new ways of putting knowledge to work, they not only attract the best people; they also make possible qualitatively new ways of competing. In "Competing on Capabilities: The New Rules of Corporate Strategy," George Stalk, Philip Evans, and Lawrence E. Shulman of The Boston Consulting Group describe how some of the most successful companies today are putting organizational innovation at the very center of their business strategy. They argue that a "fundamental shift in the logic of competition" is making such innovation more important to building sustainable competitive advantage than ever before.

According to Stalk and his colleagues, the old competitive environment was characterized by stability—durable products, unchanging customer needs, clearly defined national and regional markets. In such a world, companies occupied competitive space like squares on a chessboard, building and defending market share in clearly defined product and market segments. The key to competitive advantage was *where* a company chose to compete.

In today's business environment shaped by fragmented markets, accelerated product life cycles, and global competition, however, the old approach to strategy is ineffective. Successful competitors move quickly in and out of products, markets, and sometimes even entire businesses in a process more akin to a video game than to chess. And the key to competitive advantage is less where a company competes than *how* it competes. "The essence of strategy is *not* the structure of a company's products and markets," write Stalk and his co-authors, "but the dynamics of its behavior." And the goal of strategy is "to identify and develop the hard-to-imitate organizational capabilities that distinguish a company from its competitors in the eyes of customers." The authors' case in point is Wal-Mart, a company that in a brief ten years moved from being a small niche retailer in the South to become the largest and highest-profit retailer in the world by focusing on building core capabilities that more traditional competitors like Kmart couldn't match. The lesson for senior managers: "a CEO's success in building and managing capabilities"—in other words, his or her capacity to build the company around continuous organizational innovation—"will be the chief test of management skill in the 1990s."

In "The Knowledge-Creating Company," business theorist Ikujiro Nonaka provides a unique Japanese perspective on the practices and the roles of organizations that compete through continuous innovation. Just as many Western companies have learned from Japanese manufacturing techniques in recent years, Nonaka argues, they can also learn from the ways the best Japanese companies create new knowledge. Nonaka uses examples drawn from leading Japanese companies, such as Honda, Canon, and Matsushita, to develop his own distinctive theory of knowledge creation. His conclusion may be surprising to most Western managers.

According to Nonaka, the basic insight these companies have grasped is that creating new knowledge is not simply a matter of "processing" objective information. In fact, it is a subjective and extremely personal activity. Continuous innovation, he says, "depends on tapping the tacit and often highly subjective insights, intuitions, and hunches of individual employees." And in a comment that echoes

Drucker, a central variable in this process is the "personal commitment" of a company's employees to the organization.

Given the importance of commitment to organizational innovation, Nonaka argues that "the knowledge-creating company is as much about ideals as it is about ideas." The companies he describes think of innovation not only as a matter of inventing new business practices and building new capabilities but also as a "nonstop process of personal and organizational self-renewal." As a result, the managerial tools they use look very different from those found in most Western companies—in particular, a highly symbolic managerial language made up of "metaphors, analogies, and models." Nonaka shows how companies use these tools to create and disseminate new knowledge and to redefine the traditional roles of senior managers, middle managers, and frontline employees.

Part I of *The Learning Imperative* ends with an interview that weaves together many of the themes of the previous articles. In "Message and Muscle," Nicholas Hayek, CEO of the Swiss Corporation for Microelectronics and Watchmaking (known by the French acronym, SMH), describes how he engineered one of the world's most spectacular industrial comebacks—the revitalization of the Swiss watchmaking industry. The engine of SMH's transformation has been the Swatch Watch, one of the most successful products of recent years (in 1992 alone, SMH sold 27 *million*). According to Hayek, Swatch is a "triumph of engineering," the product of radical innovations in product design, automation, and assembly. But it is also a "triumph of the imagination," the product of human creativity and even fantasy.

Hayek's story has implications for countries as well as companies: in a knowledge economy, advanced industrial countries like Switzerland don't have to give up traditional manufacturing industries to low-wage nations. By focusing on continuous innovation, they can reinvent these industries and outmaneuver their competitors.

Designing New Behaviors in the Learning Organization

In the old days, senior managers concentrated on strategy and structure. As George Stalk and his colleagues make clear, today they must focus on strategy and *behavior*—the kind of practices and attitudes one finds at Wal-Mart, Canon, or SMH. It is relatively easy to understand how senior managers create an organization's structure. But how do they go about shaping the behaviors necessary to compete in today's more dynamic competitive environment?

The first step is to make a fundamental shift in the way senior managers think about their role. They become less "planners," "controllers," or "decision makers" and more "designers." They design the organizational context—or, in a metaphor drawn from the computer business, the "architecture"—that makes possible new ways of working and behaving. One might even say that seen from the perspective of the senior manager, a company's chief product is the organization itself. In Part II, *The Learning Imperative* turns to the ways some companies are going about designing the learning organization.

The first article describes one senior manager's effort to rethink the mission and purpose of one of America's top corporate research labs, the Xerox Corporation's prestigious Palo Alto Research Center (PARC). In "Research That Reinvents the Corporation," Xerox chief scientist and PARC director John Seely Brown explains why PARC, without giving up its traditional focus on state-of-the-art information technologies, is devoting more resources and intellectual energy to the study of the human and organizational dimensions of innovation.

One reason for this shift in emphasis is a paradox of the very computer and information technologies that PARC has done so much to pioneer. The more powerful and versatile such technologies become, the easier it becomes to use them as highly effective learning tools. Whereas in the past, people had to adapt their work practices to rigid technologies, now technology can be customized to leverage the way people actually work. As Brown puts it, "technology is finally becoming powerful enough to get out of the way."

Brown describes how he has redefined PARC's mission to improve Xerox's capacity for precisely the kind of social innovation that Peter Drucker argues is central to the modern organization. Today, PARC researchers aren't just inventing new products. They are creating new prototypes of organizational practice, helping people throughout the corporation learn from the informal local innovation that takes place continuously in any big company, and "co-producing" new business innovations with other parts of the corporation and with customers. In the process, they are redefining "innovation" to include the reinvention of the corporation itself.

One of the implications of Brown's article is that in an economy where technologies and markets are changing rapidly, competitive success is determined as much on the frontline of the company as in corporate departments. "Innovation isn't the privileged activity of the research department," he writes. "It goes on at all levels of a company." In my own article, "The Designer Organization: Italy's GFT Goes Global," I attempt to describe the efforts of one company to unleash local innovation in its units around the world and then learn from it.

Gruppo GFT is the world's largest manufacturer of designer clothing and the company behind such well-known European designer labels as Giorgio Armani, Emanuel Ungaro, and Valentino. As GFT has transformed itself from a relatively small and primarily Italian company into a billion-dollar global business, its managers have found that they have to rethink their business strategy, organizational structure, and managerial competencies and practices. Or as GFT chairman Marco Rivetti puts it, "We have to reinvent everything from scratch."

What GFT has learned is that being global isn't about standardization as much as a quantum increase in complexity. And that requires an organization that can combine the coordination of a global company with the responsiveness of small local companies, close to the market. In the words of Rivetti, "To be global means to recognize difference and be flexible enough to adapt to it." But doing this requires turning the organization inside out—and, in particular, making sure that innovations at the "periphery" of the company redirect business practices at the center. In effect, Rivetti and his managers are trying to make GFT into a "designer organization," able to adjust and adapt continuously to differences among markets and changes within markets—much as the company's designer collections change from year to year.

One popular management buzzword to describe how big organizations can develop the flexibility and responsiveness that both Xerox and GFT are aiming for is "networks." In "How Networks Reshape Organizations—for Results," management consultant Ram Charan explores how companies such as DuPont, Armstrong World Industries, Conrail, and the Royal Bank of Canada are using networks to develop an essential capability in today's economy: "superior execution in a volatile environment."

Charan makes the valuable point that these companies don't see networks as a new kind of organizational *structure*. Rather, they conceive of them as new ways of working and behaving. "Networks really begin to matter," Charan writes, "when they affect patterns of relationships and change behavior—change driven by the frequency, intensity, and honesty of the dialogue among managers on specific priorities."

But getting that kind of effective behavior in an organization requires senior managers to play a new role. Like the global managers of GFT, they must become organizational designers and create what Charan calls the new "social architecture" that allows networks to flourish. Senior managers, he says, must "become more adept at diagnosing the behavior of their organization, building relationships among key managers, modifying measures and rewards, and linking all of these 'soft' changes with the company's economic performance."

Part II of *The Learning Imperative* concludes with an interview in which Xerox CEO Paul Allaire describes how he is trying to do precisely that. Allaire's term, "organizational architecture," is slightly different from Charan's but means much the same thing. In "The CEO as Organizational Architect," Allaire describes how he and his management team have defined the new managerial behaviors necessary for Xerox to be successful in a changing competitive environment, created an "architecture" designed to encourage those behaviors, selected managers to fill the new roles within this architecture, and reshaped reward and recognition systems to fit the company's new mission and strategy.

Managing the Psychological Frontiers of the Learning Organization

More and more companies are attempting to design a learning organization. But this emphasis shouldn't make us forget Ikujiro Nonaka's central insight: innovation in a knowledge economy is frequently a profoundly personal and subjective experience. Work in a learning organization opens new psychological frontiers for all employees. And it makes the ability to reflect systematically on one's own motivations, behaviors, and work relationships a necessary part of every manager's job description.

This is uncharted territory for many managers. They feel uneasy, and perhaps even suspicious, about the more personal side of work because it takes them into the realm of the emotions, and emotions often seem too ambiguous and "soft" to be amenable to clear-cut managerial action. As Xerox CEO Paul Allaire says in his interview, "The hardest stuff is the soft stuff—values, personal style, ways of interacting."

And yet, the more competitive success depends on personal commitment, the more the "soft stuff" needs to be the object of some hard managerial attention. A central part of every manager's job is to create the shared understandings and mutual trust that make genuine collaboration among employees possible and productive. A manager's ability to build that understanding and trust depends in large part on the degree of honesty, authenticity, and personal integrity he or she brings to the job.

Part III of *The Learning Imperative* takes a first step toward providing managers with the conceptual tools they need to navigate the psychological frontiers of the learning organization. In "The New Boundaries of the 'Boundaryless' Company," Larry Hirschhorn and Thomas Gilmore of the Philadelphia consulting firm, the Center for Applied Research, provide managers with a useful map of those frontiers. Their starting point is a reflection on yet another of the popular management slogans of the 1990s: the "boundaryless company," or the idea that in order to become more flexible and innovative, organizations need to eliminate the traditional boundaries of hierarchy, function, and geography that make companies rigid and unresponsive.

When managers break down such boundaries, argue Hirschhorn and Gilmore, they often assume that this eliminates the need for boundaries altogether. But this is a serious mistake. In fact, as traditional boundaries disappear, a new set of boundaries becomes more important. These boundaries are psychological rather than organizational. As the authors put it, "They aren't drawn on a company's organizational chart but in the minds of its managers and employees."

Hirschhorn and Gilmore describe some of the challenges of managing these psychological boundaries: how to give up control without abdicating authority, how to combat the myopia of traditional functional perspectives without losing the advantages of specialization, how to encourage workers to identify with their local work groups without undermining their connection to the organization as a whole, and how to encourage constructive conflict that surfaces and resolves legitimate differences without tearing the organization apart.

In developing these new psychological skills, say Hirschhorn and Gilmore, managers have an unexpected resource: their own gut feelings about work and the people with whom they do it. Strong feelings, positive or negative, aren't just the inevitable emotional fallout of work relationships; they are valuable *data*—clues to the dynamics of a manager's boundary relationships. "In this respect," they write, "feelings are an aid to thinking and to managing, they are a real part of real work."

Hirschhorn and Gilmore emphasize how a manager's feelings can be an indispensable learning tool for managerial thinking and behavior. In "Teaching Smart People How to Learn," Harvard Business School professor Chris Argyris reverses the equation. His focus is on the ways that a manager's style of thinking can distort feelings and behavior and obstruct learning.

Argyris points out that the learning imperative is too often a "learning dilemma." "Success in the marketplace increasingly depends on learning, yet most people don't know how to learn." What's more, he argues, the people whom many assume to be the best at learning are in fact the same people who have some of the greatest difficulty with it: the well-educated, high-commitment professionals who occupy leadership positions in the modern corporation.

According to Argyris, professionals are often extremely adept at "problem solving," that is, analyzing and correcting errors in the external environment. But while solving problems is important, continuous learning also depends on managers' ability to look inward, "to reflect critically on their own behavior, identify the ways they often inadvertently contribute to the organization's problems, and then change how they act."

Paradoxically, the reason that many professionals are so bad at learning is that they are so good at their work. "Because many professionals are almost always successful at what they do," Argyris writes, "they rarely experience failure. And because they have rarely failed, they have never learned how to learn from failure." As a result, they find it difficult to reflect on their own behavior—and their possible errors—in an open, nondefensive way.

Such defensive behavior is pervasive in organizations, says Argyris. But companies can learn how to deal with it. The solution is to focus not on professionals' motivations or intentions but on the way they *reason* about their own behavior—that is, the cognitive rules they use to design and then implement their actions. Argyris makes the valuable point that navigating the psychological frontiers of the learning organization has nothing "touchy-feely" or "soft" about it. Indeed, managers can develop rules for "tough reasoning" about their own behavior as analytical and "hard" as the most sophisticated ideas in strategy or finance.

Many of the themes raised by Hirschhorn and Gilmore and by Argyris are echoed in the CEO interview which closes Part III. In "Values Make the Company," Levi Strauss CEO Robert Haas describes his efforts to use values as a way to tap into the more personal dimension of work in the new economy. For Haas, new competitive realities guarantee that the "hard stuff" and the "soft stuff" of management are inextricably linked. To react quickly to changes in markets and technology, companies need to put more accountability, authority, and information into the hands of people on the front lines of the business. Doing so, however, also requires an enormous diffusion of power. "The controls have to be conceptual," says Haas. "It's the *ideas* of a business that are controlling, not some manager with authority. Values provide a common language for aligning a company's leadership and its people."

Haas describes how Levi is using the values articulated in the company's Aspirations Statement to teach managers how to examine their own behavior critically, set broad parameters for employees without abdicating authority, and encourage disagreement and productive dialogue. He also describes how the company's values shape everything from the way Levi Strauss defines occupational roles and responsibilities to how it conducts performance evaluations, trains new employees, organizes work, and makes business decisions. And he acknowledges the challenges and the difficulties of making values real not just for managers at the top but for employees throughout the organization.

Getting from Here to There

Of course, some parts of the manager's job have *not* changed with the arrival of a new economy. One of them is the manager's traditional bias toward action. It's one thing for managers to understand that in an economy built on continuous innovation, learning is an imperative. It is quite another to commit to building a learning organization and to stick to that commitment in the heat of the day-to-day demands and distractions of managing large organizations. For this reason, *The Learning Imperative* ends with four profiles that emphasize how companies can get "from here to there."

Part IV begins with an article that offers a valuable lesson in how *not* to go about creating an innovative organization. In "Why Change Programs Don't Produce Change," Michael Beer of the Harvard Business School, Russell A. Eisenstat, an independent consultant, and Bert Spector of Northeastern University's College of Business Administration argue that while most senior managers understand the need for change, they fundamentally misunderstand how to bring it about. In particular, they make two mistakes: they assume that change must come "from the top down" through companywide programs; and they think that the chief component of any successful change program is the reorganization of the company's formal structure and systems.

A four-year study of organizational change at six large corporations convinced Beer and his colleagues that both assumptions are dangerous. In fact, successful change efforts rarely start at the top. Far more often, they begin at the periphery of the company in a few plants or divisions far from corporate headquarters. And these local initiatives focus, not on changing formal structures and systems, but on creating ad-hoc organizational arrangements to solve concrete business problems. Beer and his co-authors describe a six-step "critical path" for corporate renewal, using as their example the change process at a division of a major U.S. corporation facing the task of commercializing a product originally designed for the military market.

As Beer, Eisenstat, and Spector suggest, the path to continuous organizational innovation is rarely simple and clear-cut. Consider, for example, how Motorola went about expanding its $7-million annual training budget into a $120-million annual investment in education. According to William Wiggenhorn, Motorola's corporate vice president for training and education and author of "Motorola U.: When Training Becomes an Education," it was an "odyssey, a ten-year expedition full of well-meaning mistakes, heroic misapprehensions, and shocking discoveries."

Wiggenhorn's article (which won the 1990 McKinsey Award) is a dramatic portrait of the factors that pushed one of America's best companies to learn just how important learning is. It describes the thought process that led Motorola's top managers to the remarkable conclusion that "successful companies in today's business climate must not only train workers but build educational systems."

From the story of the establishment of Motorola's educational system, Part IV turns to a manufacturing plant where the production system itself is the educational system. Proponents of the learning organization often assume that bureaucracy is bad for learning. Standardized work and fragmented tasks rob work of its meaning and kill employee motivation and creativity—or so the usual argument goes. In "Time-and-Motion Regained" University of Southern California business professor Paul S. Adler makes the provocative claim that the usual argument has it all wrong. Standardization and specialization, properly understood and organized, can be a tremendous stimulus to learning. It can result in what Adler calls a "learning bureaucracy."

Adler's case is the Fremont, California, plant of New United Motor Manufacturing, Inc., a joint venture of General Motors and Toyota designed to apply Japanese manufacturing techniques at a U.S. auto plant. When run by GM, the Fremont facility was what one manager described as "the worst plant in the world," with the lowest productivity in the GM system, abysmal quality, and rampant drug and alcohol abuse among employees. Using largely the same work force (at its opening, 85% of NUMMI's 2,200 hourly workers had previously worked at the Fremont plant), the joint venture has reached productivity and quality levels higher than those of any other GM facility and nearly as high as Toyota's Japanese plants. What's more, absenteeism has plummeted, and NUMMI's employees report relatively high levels of job satisfaction.

According to Adler, the combination of Toyota's standardized production system with a social context of labor-management cooperation and active union and worker input into job design and standards produces a highly standardized work environment that nevertheless stimulates worker learning. At NUMMI, "standardization is the essential precondition of learning." Both Adler's argument and his case study provide a remarkably clear-eyed view of work and motivation in a state-of-the-art manufacturing plant.

Part IV of *The Learning Imperative* ends with an interview with Arden C. Sims, CEO of Globe Metallurgical, the first small company to win the coveted Malcolm Baldrige National Quality Award. In "Trial-by-Fire Transformation," Sims recounts in vivid detail Globe's change from a sluggish supplier of commodity metals to the ailing steel industry into the preeminent supplier of specialty metals to the chemical and foundry industries around the world.

Sims reveals that Globe's extraordinary turnaround was not the result of hiring high-paid consultants, listening to the latest management gurus, or, for that matter, reading articles like those that appear in this book. Rather, it was the product of managers squarely confronting complex business crises and, through a process of trial and error, figuring out what worked. The result was one that any company would be proud of: organizational innovations—a management-led leveraged buyout, flexible work teams, a new strategic focus on R&D and high-value-added niche marketing—that make for competitive success in the new economy.

From Peter Drucker's theories of the knowledge economy to Arden Sims' description of the gritty realities of global competition in a small manufacturing company, *The Learning Imperative* is about management as innovation. What I have tried to make clear in this introduction is the way each article in this volume is enhanced by the opportunity for "dialogue" with every other. Taken individually, these articles are important contributions to the managerial debate about organizational learning. Taken together, they are an invaluable resource for any manager striving to cope with the challenge of organizational innovation and change.

The articles collected in this volume were the product of a highly collaborative work environment. For that reason, I would like to end this introduction by thanking all my former colleagues at the *Harvard Business Review*. Without their willingness to read and react to manuscripts, discuss ideas, offer criticism, and share advice, this book would simply not exist. For a time, they made HBR a stimulating and remarkably innovative place to work and to learn.

Because of the key role they played on specific articles appearing in this book, some former and current HBR editors deserve special mention by name. Bernard Avishai, Nan Stone, William Taylor, and Thomas Teal developed or edited one or more of the contributions in this volume. Finally, Alan M. Webber was an indispensable intellectual partner and influence on nearly all of them.

The
Learning
Imperative

PART

I

Decoding the Business Logic

1
The New Society of Organizations

Peter F. Drucker

Every few hundred years throughout Western history, a sharp trans-formation has occurred. In a matter of decades, society altogether rearranges itself—its world view, its basic values, its social and political structures, its arts, its key institutions. Fifty years later a new world exists. And the people born into that world cannot even imagine the world in which their grandparents lived and into which their own parents were born.

Our age is such a period of transformation. Only this time the transformation is not confined to Western society and Western history. Indeed, one of the fundamental changes is that there is no longer a "Western" history or a "Western" civilization. There is only world history and world civilization.

Whether this transformation began with the emergence of the first non-Western country, Japan, as a great economic power or with the first computer—that is, with information—is moot. My own candidate would be the GI Bill of Rights, which gave every American soldier returning from World War II the money to attend a university, some-thing that would have made absolutely no sense only 30 years earlier at the end of World War I. The GI Bill of Rights and the enthusiastic response to it on the part of America's veterans signaled the shift to a knowledge society.

In this society, knowledge is *the* primary resource for individuals and for the economy overall. Land, labor, and capital—the economist's traditional factors of production—do not disappear, but they become secondary. They can be obtained, and obtained easily, provided there is specialized knowledge. At the same time, however, specialized knowledge by itself produces nothing. It can become productive only when it is integrated into a task. And that is why the knowledge society is also a society of organizations: the purpose and function of every organization, business and nonbusiness alike, is the integration of specialized knowledges into a common task.

If history is any guide, this transformation will not be completed until 2010 or 2020. Therefore, it is risky to try to foresee in every detail the world that is emerging. But what new questions will arise and where the big issues will lie we can, I believe, already discover with a high degree of probability.

In particular, we already know the central tensions and issues that confront the society of organizations: the tension created by the community's need for stability and the organization's need to destabilize; the relationship between individual and organization and the responsibilities of one to another; the tension that arises from the organization's need for autonomy and society's stake in the Common Good; the rising demand for socially responsible organizations; the tension between specialists with specialized knowledges and performance as a team. All of these will be central concerns, especially in the developed world, for years to come. They will not be resolved by pronunciamento or philosophy or legislation. They will be resolved where they originate: in the individual organization and in the manager's office.

Society, community, and family are all conserving institutions. They try to maintain stability and to prevent, or at least to slow, change. But the modern organization is a destabilizer. It must be organized for innovation, and innovation, as the great Austro-American economist Joseph Schumpeter said, is "creative destruction." And it must be organized for the systematic abandonment of whatever is established, customary, familiar, and comfortable, whether that is a product, service, or process; a set of skills; human and social relationships; or the organization itself. In short, it must be organized for constant change. The organization's function is to put knowledge to work—on tools, products, and processes; on the design of work; on knowledge itself. It is the nature of knowledge that it changes fast and that today's certainties always become tomorrow's absurdities.

Skills change slowly and infrequently. If an ancient Greek stonecutter came back to life today and went to work in a stone mason's yard, the only change of significance would be the design he was asked to carve on the tombstones. The tools he would use are the same, only now they have electric batteries in the handles. Throughout history, the craftsman who had learned a trade after five or seven years of apprenticeship had learned, by age eighteen or nineteen, everything he would ever need to use during his lifetime. In the society of organizations, however, it is safe to assume that anyone with any knowledge will have to acquire new knowledge every four or five years or become obsolete.

This is doubly important because the changes that affect a body of knowledge most profoundly do not, as a rule, come out of its own domain. After Gutenberg first used movable type, there was practically no change in the craft of printing for 400 years—until the steam engine came in. The greatest challenge to the railroad came not from changes in railroading but from the automobile, the truck, and the airplane. The pharmaceutical industry is being profoundly changed today by knowledge coming from genetics and microbiology, disciplines that few biologists had heard of 40 years ago.

And it is by no means only science or technology that creates new knowledge and makes old knowledge obsolete. Social innovation is equally important and often more important than scientific innovation. Indeed, what triggered the present worldwide crisis in that proudest of nineteenth-century institutions, the commercial bank, was not the computer or any other technological change. It was the discovery by nonbankers that an old but hitherto rather obscure financial instrument, commercial paper, could be used to finance companies and would thus deprive the banks of the business on which they had held a monopoly for 200 years and which gave them most of their income: the commercial loan. The greatest change of all is probably that in the last 40 years purposeful innovation—both technical and social—has itself become an organized discipline that is both teachable and learnable.

Nor is rapid knowledge-based change confined to business, as many still believe. No organization in the 50 years since World War II has changed more than the U.S. military. Uniforms have remained the same. Titles of rank have remained the same. But weapons have changed completely, as the Gulf War of 1991 dramatically demonstrated; military doctrines and concepts have changed even more drastically, as have the armed services' organizational structures, command structures, relationships, and responsibilities.

Similarly, it is a safe prediction that in the next 50 years, schools and universities will change more and more drastically than they have since they assumed their present form more than 300 years ago when they reorganized themselves around the printed book. What will force these changes is, in part, new technology, such as computers, videos, and telecasts via satellite; in part the demands of a knowledge-based society in which organized learning must become a lifelong process for knowledge workers; and in part new theory about how human beings learn.

For managers, the dynamics of knowledge impose one clear imperative: every organization has to build the management of change into its very structure.

On the one hand, this means every organization has to prepare for the abandonment of everything it does. Managers have to learn to ask every few years of every process, every product, every procedure, every policy: "If we did not do this already, would we go into it now knowing what we now know?" If the answer is no, the organization has to ask, "So what do we do now?" And it has to *do* something, and not say, "Let's make another study." Indeed, organizations increasingly will have to *plan* abandonment rather than try to prolong the life of a successful product, policy, or practice—something that so far only a few large Japanese companies have faced up to.

On the other hand, every organization must devote itself to creating the new. Specifically, every management has to draw on three systematic practices. The first is continuing improvement of everything the organization does, the process the Japanese call *kaizen*. Every artist throughout history has practiced kaizen, or organized, continuous self-improvement. But so far only the Japanese—perhaps because of their Zen tradition—have embodied it in the daily life and work of their business organizations (although not in their singularly change-resistant universities). The aim of kaizen is to improve a product or service so that it becomes a truly different product or service in two or three years' time.

Second, every organization will have to learn to exploit its knowledge, that is, to develop the next generation of applications from its own successes. Again, Japanese businesses have done the best with this endeavor so far, as demonstrated by the success of the consumer electronics manufacturers in developing one new product after another from the same American invention, the tape recorder. But successful exploitation of their successes is also one of the strengths of the fast-growing American pastoral churches.

Finally, every organization will have to learn to innovate—and innovation can now be organized and must be organized—as a systematic process. And then, of course, one comes back to abandonment, and the process starts all over. Unless this is done, the knowledge-based organization will very soon find itself obsolescent, losing performance capacity and with it the ability to attract and hold the skilled and knowledgeable people on whom its performance depends.

The need to organize for change also requires a high degree of decentralization. That is because the organization must be structured to make decisions quickly. And those decisions must be based on closeness—to performance, to the market, to technology, and to all the many changes in society, the environment, demographics, and knowledge that provide opportunities for innovation if they are seen and utilized.

All this implies, however, that the organizations of the post-capitalist society must constantly upset, disorganize, and destabilize the community. They must change the demand for skills and knowledges: just when every technical university is geared up to teach physics, organizations need geneticists. Just when bank employees are most proficient in credit analysis, they will need to be investment counselors. But also, businesses must be free to close factories on which local communities depend for employment or to replace grizzled model makers who have spent years learning their craft with 25-year-old whiz kids who know computer simulation.

Similarly, hospitals must be able to move the delivery of babies into a free-standing birthing center when the knowledge base and technology of obstetrics change. And we must be able to close a hospital altogether when changes in medical knowledge, technology, and practice make a hospital with fewer than 200 beds both uneconomical and incapable of giving first-rate care. For a hospital—or a school or any other community organization—to discharge its social function we must be able to close it down, no matter how deeply rooted in the local community it is and how much beloved, if changes in demographics, technology, or knowledge set new prerequisites for performance.

But every one of such changes upsets the community, disrupts it, deprives it of continuity. Every one is "unfair." Every one destabilizes.

Equally disruptive is another fact of organizational life: the modern organization must be *in* a community but cannot be *of* it. An organization's members live in a particular place, speak its language, send their children to its schools, vote, pay taxes, and need to feel at home there. Yet the organization cannot submerge itself in the community nor subordinate itself to the community's ends. Its "culture" has to transcend community.

It is the nature of the task, not the community in which the task is being performed, that determines the culture of an organization. The American civil servant, though totally opposed to communism, will understand immediately what a Chinese colleague tells him about bureaucratic intrigues in Beijing. But he would be totally baffled in his own Washington, D.C. if he were to sit in on a discussion of the next week's advertising promotions by the managers of the local grocery chain.

To perform its task the organization has to be organized and managed the same way as others of its type. For example, we hear a great deal about the differences in management between Japanese and American companies. But a large Japanese company functions very much like a large American company; and both function very much like a large German or British company. Likewise, no one will ever doubt that he or she is in a hospital, no matter where the hospital is located. The same holds true for schools and universities, for labor unions and research labs, for museums and opera houses, for astronomical observatories and large farms.

In addition, each organization has a value system that is determined by its task. In every hospital in the world, health care is considered the ultimate good. In every school in the world, learning is considered the ultimate good. In every business in the world, production and distribution of goods or services are considered the ultimate good. For the organization to perform to a high standard, its members must believe that what it is doing is, in the last analysis, the one contribution to community and society on which all others depend.

In its culture, therefore, the organization will always transcend the community. If an organization's culture and the values of its community clash, the organization must prevail—or else it will not make its social contribution. "Knowledge knows no boundaries," says an old proverb. There has been a "town and gown" conflict ever since the first university was established more than 750 years ago. But such a conflict—between the autonomy the organization needs in order to perform and the claims of the community, between the values of the organization and those of the community, between the decisions facing the organization and the interests of the community—is inherent in the society of organizations.

The issue of social responsibility is also inherent in the society of organizations. The modern organization has and must have social power—and a good deal of it. It needs power to make decisions about people: whom to hire, whom to fire, whom to promote. It needs power to establish the rules and disciplines required to produce results: for example, the assignment of jobs and tasks and the establishment of working hours. It needs power to decide which factories to build where and which factories to close. It needs power to set prices, and so on.

And nonbusinesses have the greatest social power—far more, in fact, than business enterprises. Few organizations in history were ever granted the power the university has today. Refusing to admit a student or to grant a student the diploma is tantamount to debarring that person from careers and opportunities. Similarly, the power of the American hospital to deny a physician admitting privileges is the power to exclude that physician from the practice of medicine. The labor union's power over admission to apprenticeship or its control of access to employment in a "closed shop," where only union members can be hired, gives the union tremendous social power.

The power of the organization can be restrained by political power. It can be made subject to due process and to review by the courts. But it must be exercised by individual organizations rather than by political authorities. This is why post-capitalist society talks so much about social responsibilities of the organization.

It is futile to argue, as Milton Friedman, the American economist and Nobel-laureate does, that a business has only one responsibility: economic performance. Economic performance is the *first* responsibility of a business. Indeed, a business that does not show a profit at least equal to its cost of capital is irresponsible; it wastes society's resources. Economic performance is the base without which a business cannot discharge any other responsibilities, cannot be a good employer, a good citizen, a good neighbor. But economic performance is not the *only* responsibility of a business any more than educational performance is the only responsibility of a school or health care the only responsibility of a hospital.

Unless power is balanced by responsibility, it becomes tyranny. Furthermore, without responsibility power always degenerates into nonperformance, and organizations must perform. So the demand for socially responsible organizations will not go away but rather widen.

Fortunately, we also know, if only in rough outline, how to answer the problem of social responsibility. Every organization must assume full responsibility for its impact on employees, the environment, customers, and whomever and whatever it touches. That is its social responsibility. But we also know that society will increasingly look to major organizations, for-profit and nonprofit alike, to tackle major social ills. And there we had better be watchful because good intentions are not always socially responsible. It is irresponsible for an organization to accept—let alone to pursue—responsibilities that would impede its capacity to perform its main task and mission or to act where it has no competence.

Organization has become an everyday term. Everybody nods when somebody says, "In our organization, everything should revolve around the customer" or "In this organization, they never forget a mistake." And most, if not all, social tasks in every developed country are performed in and by an organization of one kind or another. Yet no one in the United States—or anyplace else—talked of "organizations" until after World War II. *The Concise Oxford Dictionary* did not even list the term in its current meaning in the 1950 edition. It is only the emergence of management since World War II, what I call the "Management Revolution," that has allowed us to see that the organization is discrete and distinct from society's other institutions.

Unlike "community," "society," or "family," organizations are purposefully designed and always specialized. Community and society are defined by the bonds that hold their members together, whether they be language, culture, history, or locality. An organization is defined by its task. The symphony orchestra does not attempt to cure the sick; it plays music. The hospital takes care of the sick but does not attempt to play Beethoven.

Indeed, an organization is effective only if it concentrates on one task. Diversification destroys the performance capacity of an organization, whether it is a business, a labor union, a school, a hospital, a community service, or a house of worship. Society and community must be multidimensional; they are environments. An organization is a tool. And as with any other tool, the more specialized it is, the greater its capacity to perform its given task.

Because the modern organization is composed of specialists, each with his or her own narrow area of expertise, its mission must be crystal clear. The organization must be single-minded, or its members will become confused. They will follow their own specialty rather than apply it to the common task. They will each define "results" in terms of their own specialty and impose its values on the organization. Only a focused and common mission will hold the organization together and enable it to produce. Without such a mission, the organization will soon lose credibility and, with it, its ability to attract the very people it needs to perform.

It can be all too easy for managers to forget that joining an organization is always voluntary. De facto there may be little choice. But even where membership is all but compulsory—as membership in the Catholic church was in all the countries of Europe for many centuries for all but a handful of Jews and Gypsies—the fiction of voluntary choice is always carefully maintained: the godfather at the infant's baptism pledges the child's voluntary acceptance of membership in the church.

Likewise, it may be difficult to leave an organization—the Mafia, for instance, a big Japanese company, the Jesuit order. But it is always possible. And the more an organization becomes an organization of knowledge workers, the easier it is to leave it and move elsewhere. Therefore, an organization is always in competition for its most essential resource: qualified, knowledgeable people.

All organizations now say routinely, "People are our greatest asset." Yet few practice what they preach, let alone truly believe it. Most still believe, though perhaps not consciously, what nineteenth-century employers believed: people need us more than we need them. But, in fact, organizations have to market membership as much as they market products and services—and perhaps more. They have to attract people, hold people, recognize and reward people, motivate people, and serve and satisfy people.

The relationship between knowledge workers and their organizations is a distinctly new phenomenon, one for which we have no good term. For example, an employee, by definition, is someone who gets paid for working. Yet the largest single group of "employees" in the United States is comprised of the millions of men and women who work several hours a week without pay for one or another nonprofit organization. They are clearly "staff" and consider themselves as such, but they are unpaid volunteers. Similarly, many people who work as employees are not employed in any legal sense because they do not work for someone else. Fifty or sixty years ago, we would have spoken of these people (many, if not most, of whom are educated professionals) as "independent"; today we speak of the "self-employed."

These discrepancies—and they exist in just about every language—remind us why new realities often demand new words. But until such a word emerges, this is probably the best definition of employees in the post-capitalist society: people whose ability to make a contribution depends on having access to an organization.

As far as the employees who work in subordinate and menial occupations are concerned—the salesclerk in the supermarket, the cleaning woman in the hospital, the delivery-truck driver—the consequences of this new definition are small. For all practical purposes, their position may not be too different from that of the wage earner, the "worker" of yesterday, whose direct descendants they are. In fact, this is precisely one of the central social problems modern society faces.

But the relationship between the organization and knowledge workers, who already number at least one-third and more likely two-fifths of all employees, is radically different, as is that between the organization and volunteers. They can work only because there is an organization, thus they too are dependent. But at the same time, they own the "means of production"—their knowledge. In this respect, they are independent and highly mobile.

Knowledge workers still need the tools of production. In fact, capital investment in the tools of the knowledge employee may already be higher than the capital investment in the tools of the manufacturing worker ever was. (And the social investment, for example, the investment in a knowledge worker's education, is many times the investment in the manual worker's education.) But this capital investment is unproductive unless the knowledge worker brings to bear on it the knowledge that he or she owns and that cannot be taken away. Machine operators in the factory did as they were told. The machine decided not only what to do but how to do it. The knowledge employee may well need a machine, whether it be a computer, an ultrasound analyzer, or a telescope. But the machine will not tell the knowledge worker what to do, let alone how to do it. And without this knowledge, which belongs to the employee, the machine is unproductive.

Further, machine operators, like all workers throughout history, could be told what to do, how to do it, and how fast to do it. Knowledge workers cannot be supervised effectively. Unless they know more about their specialty than anybody else in the organization, they are basically useless. The marketing manager may tell the market researcher what the company needs to know about the design of a new product and the market segment in which it should be positioned. But it is the market researcher's job to tell the president of the company what market research is needed, how to set it up, and what the results mean.

During the traumatic restructuring of American business in the 1980s, thousands, if not hundreds of thousands, of knowledge employees lost their jobs. Their companies were acquired, merged, spun off, or liquidated. Yet within a few months, most of them found new jobs in which to put their knowledge to work. The transition period was painful, and in about half the cases, the new job did not pay quite as much as the old one did and may not have been as enjoyable. But the laid-off technicians, professionals, and managers found they had the "capital," the knowledge: they owned the means of production. Somebody else, the organization, had the tools of production. The two needed each other.

One consequence of this new relationship—and it is another new tension in modern society—is that loyalty can no longer be obtained by the paycheck. The organization must earn loyalty by proving to its knowledge employees that it offers them exceptional opportunities for putting their knowledge to work. Not so long ago we talked about "labor." Increasingly we are talking about "human resources." This change reminds us that it is the individual, and especially the skilled and knowledgeable employee, who decides in large measure what he or she will contribute to the organization and how great the yield from his or her knowledge will be.

Because the modern organization consists of knowledge specialists, it has to be an organization of equals, of colleagues and associates. No knowledge ranks higher than another; each is judged by its contribution to the common task rather than by any inherent superiority or inferiority. Therefore, the modern organization cannot be an organization of boss and subordinate. It must be organized as a team.

There are only three kinds of teams. One is the sort of team that plays together in tennis doubles. In that team—and it has to be small—each member adapts himself or herself to the personality, the skills, the strengths, and the weaknesses of the other member or members. Then there is the team that plays European football or soccer. Each player has a fixed position; but the whole team moves together (except for the goalie) while individual members retain their relative positions. Finally, there is the American baseball team—or the orchestra—in which all the members have fixed positions.

At any given time, an organization can play only one kind of game. And it can use only one kind of team for any given task. Which team to use or game to play is one of the riskiest decisions in the life of an organization. Few things are as difficult in an organization as transforming from one kind of team to another.

Traditionally, American industry used a baseball-style team to produce a new product or model. Research did its work and passed it on to engineering. Engineering did its work and passed it on to manufacturing. Manufacturing did its work and passed it on to marketing. Accounting usually came in at the manufacturing phase. Personnel usually came in only when there was a true crisis—and often not even then.

Then the Japanese reorganized their new product development into a soccer team. In such a team, each function does its own work, but from the beginning they work together. They move with the task, so to speak, the way a soccer team moves with the ball. It took the Japanese at least 15 years to learn how to do this. But once they had mastered the new concept, they cut development time by two-thirds. Where traditionally it has taken 5 years to bring out a new automobile model, Toyota, Nissan, and Honda now do it in 18 months. This, as much as their quality control, has given the Japanese the upper hand in both the American and European automobile markets.

Some American manufacturers have been working hard to reorganize their development work according to the Japanese model. Ford Motor Company, for instance, began to do so in the early 1980s. Ten years later, in the early 1990s, it has made considerable progress—but not nearly enough to catch up with the Japanese. Changing a team demands the most difficult learning imaginable: unlearning. It demands giving up hard-earned skills, habits of a lifetime, deeply cherished values of craftsmanship and professionalism, and—perhaps the most difficult of all—it demands giving up old and treasured human relationships. It means abandoning what people have always considered "our community" or "our family."

But if the organization is to perform, it must be organized as a team. When modern organizations first arose in the closing years of the nineteenth century, the only model was the military. The Prussian Army was as much a marvel of organization for the world of 1870 as Henry Ford's assembly line was for the world of 1920. In the army of 1870, each member did much the same thing, and the number of people with any knowledge was infinitesimally small. The army was organized by command-and-control, and business enterprise as well as most other institutions copied that model. This is now rapidly changing. As more and more organizations become information-based, they are transforming themselves into soccer or tennis teams, that is, into responsibility-based organizations in which every member must act as a responsible decision maker. All members, in other words, have to see themselves as "executives."

Even so, an organization must be managed. The management may be intermittent and perfunctory, as it is, for instance, in the Parent-Teacher Association at a U.S. suburban school. Or management may be a full-time and demanding job for a fairly large group of people, as it is in the military, the business enterprise, the labor union, and the university. But there have to be people who make decisions or nothing will ever get done. There have to be people who are accountable for the organization's mission, its spirit, its performance, its results. Society, community, and family may have "leaders," but only organizations know a "management." And while this management must have considerable authority, its job in the modern organization is not to command. It is to inspire.

The society of organizations is unprecedented in human history. It is unprecedented in its performance capacity both because each of its constituent organizations is a highly specialized tool designed for one specific task and because each bases itself on the organization and deployment of knowledge. It is unprecedented in its structure. But it is also unprecedented in its tensions and problems. Not all of these are serious. In fact, some of them we already know how to resolve—issues of social responsibility, for example. But there are other areas where we do not know the right answer and where we may not even be asking the right questions yet.

There is, for instance, the tension between the community's need for continuity and stability and the organization's need to be an innovator and destabilizer. There is the split between "literati" and "managers." Both are needed: the former to produce knowledge, the latter to apply knowledge and make it productive. But the former focus on words and ideas, the latter on people, work, and performance. There is the threat to the very basis of the society of organizations—the knowledge base—that arises from ever greater specialization, from the shift from knowledge to *knowledges*. But the greatest and most difficult challenge is that presented by society's new pluralism.

For more than 600 years, no society has had as many centers of power as the society in which we now live. The Middle Ages indeed knew pluralism. Society was composed of hundreds of competing and autonomous power centers: feudal lords and knights, exempt bishoprics, autonomous monasteries, "free" cities. In some places, the Austrian Tyrol, for example, there were even "free peasants," beholden to no one but the Emperor. There were also autonomous craft guilds and transnational trading leagues like the Hanseatic Merchants and the merchant bankers of Florence, toll and tax collectors, local "parliaments" with legislative and tax-raising powers, private armies available for hire, and myriads more.

Modern history in Europe—and equally in Japan—has been the history of the subjugation of all competing centers of power by one central authority, first called the "prince," then the "state." By the middle of the nineteenth century, the unitary state had triumphed in every developed country except the United States, which remained profoundly pluralistic in its religious and educational organizations. Indeed, the abolition of pluralism was the "progressive" cause for nearly 600 years.

But just when the triumph of the state seemed assured, the first new organization arose—the large business enterprise. (This, of course, always happens when the "End of History" is announced.) Since then, one new organization after another has sprung up. And old organizations like the university, which in Europe seemed to have been brought safely under the control of central governments, have become autonomous again. Ironically, twentieth-century totalitarianism, especially communism, represented the last desperate attempt to save the old progressive creed in which there is only one center of power and one organization rather than a pluralism of competing and autonomous organizations.

That attempt failed, as we know. But the failure of central authority, in and of itself, does nothing to address the issues that follow from a pluralistic society. To illustrate, consider a story that many people have heard or, more accurately, misheard.

During his lifetime, Charles E. Wilson was a prominent personality in the United States, first as president and chief executive officer of General Motors, at that time the world's largest and most successful manufacturer, then as secretary of defense in the Eisenhower administration. But if Wilson is remembered at all today it is for something he did *not* say: "What is good for General Motors is good for the United States." What Wilson actually said in his 1953 confirmation hearings for the Defense Department job was: "What is good for the United States is good for General Motors."

Wilson tried for the remainder of his life to correct the misquote. But no one listened to him. Everyone argued, "If he didn't say it, he surely believes it—in fact he *should* believe it." For as has been said, executives in an organization—whether business or university or hospital or the Boy Scouts—must believe that its mission and task are society's most important mission and task as well as the foundation for everything else. If they do not believe this, their organization will soon lose faith in itself, self-confidence, pride, and the ability to perform.

The diversity that is characteristic of a developed society and that provides its great strength is only possible because of the specialized, single-task organizations that we have developed since the Industrial Revolution and, especially, during the last 50 years. But the feature that gives them the capacity to perform is precisely that each is autonomous and specialized, informed only by its own narrow mission and vision, its own narrow values, and not by any consideration of society and community.

Therefore, we come back to the old—and never resolved—problem of the pluralistic society: Who takes care of the Common Good? Who defines it? Who balances the separate and often competing goals and values of society's institutions? Who makes the trade-off decisions and on what basis should they be made?

Medieval feudalism was replaced by the unitary sovereign state precisely because it could not answer these questions. But the unitary sovereign state has now itself been replaced by a new pluralism—a pluralism of function rather than one of political power—because it could neither satisfy the needs of society nor perform the necessary tasks of community. That, in the final analysis, is the most fundamental lesson to be learned from the failure of socialism, the failure of the belief in the all-embracing and all-powerful state. The challenge that faces us now, and especially in the developed, free-market democracies such as the United States, is to make the pluralism of autonomous, knowledge-based organizations redound both to economic performance and to political and social cohesion.

2

Competing on Capabilities: The New Rules of Corporate Strategy

George Stalk, Jr., Philip Evans, and Lawrence E. Shulman

In the 1980s, companies discovered time as a new source of competitive advantage. In the 1990s, they will learn that time is just one piece of a more far-reaching transformation in the logic of competition.

Companies that compete effectively on time—speeding new products to market, manufacturing just in time, or responding promptly to customer complaints—tend to be good at other things as well: for instance, the consistency of their product quality, the acuity of their insight into evolving customer needs, the ability to exploit emerging markets, enter new businesses, or generate new ideas and incorporate them in innovations. But all these qualities are mere reflections of a more fundamental characteristic: a new conception of corporate strategy that we call "capabilities-based competition."

For a glimpse of the new world of capabilities-based competition, consider the astonishing reversal of fortunes represented by Kmart and Wal-Mart:

In 1979, Kmart was king of the discount retailing industry, an industry it had virtually created. With 1,891 stores and average revenues per store of $7.25 million, Kmart enjoyed enormous size advantages. This allowed economies of scale in purchasing, distribution, and marketing that, according to just about any management textbook, are crucial to competitive success in a mature and low-growth industry. By contrast, Wal-Mart was a small niche retailer in the South with only 229 stores and average revenues about half of those of Kmart stores—hardly a serious competitor.

And yet, only ten years later, Wal-Mart had transformed itself and the discount retailing industry. Growing nearly 25% a year, the company achieved the highest sales per square foot, inventory turns, and operating profit of any discount retailer. Its 1989 pretax return on sales was 8%, nearly double that of Kmart. (See Exhibit I.)

Today Wal-Mart is the largest and highest profit retailer in the world—a performance that has translated into a 32% return on equity and a market valuation more than ten times book value. What's more, Wal-Mart's growth has been concentrated in half the United States, leaving ample room for further expansion. If Wal-Mart continues to gain market share at just one-half its historical rate, by 1995 the company will have eliminated all competitors from discount retailing with the exception of Kmart and Target.

The Secret of Wal-Mart's Success

What accounts for Wal-Mart's remarkable success? Most explanations focus on a few familiar and highly visible factors: the genius of founder Sam Walton, who inspires his employees and has molded a culture of service excellence; the "greeters" who welcome customers at the door; the motivational power of allowing employees to own part of the business; the strategy of "everyday low prices" that offers the customer a better deal and saves on merchandising and advertising costs. Economists also point to Wal-Mart's big stores, which offer economies of scale and a wider choice of merchandise.

But such explanations only redefine the question. *Why* is Wal-Mart able to justify building bigger stores? Why does Wal-Mart alone have a cost structure low enough to accommodate everyday low prices and greeters? And what has enabled the company to continue to grow far beyond the direct reach of Sam Walton's magnetic personality? The real secret of Wal-Mart's success lies deeper, in a set of strategic business decisions that transformed the company into a capabilities-based competitor.

The starting point was a relentless focus on satisfying customer needs. Wal-Mart's goals were simple to define but hard to execute: to provide customers access to quality goods, to make these goods available when and where customers want them, to develop a cost structure that enables competitive pricing, and to build and maintain a reputation for absolute trustworthiness. The key to achieving these goals was to make the way the company replenished inventory the centerpiece of its competitive strategy.

Exhibit I.

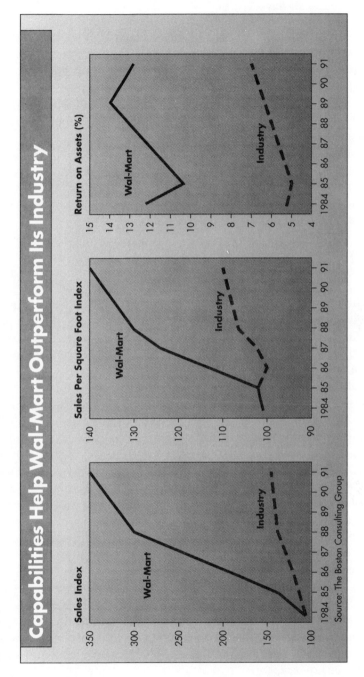

Capabilities Help Wal-Mart Outperform Its Industry

Sales Index

Wal-Mart

Industry

350
300
250
200
150
100

1984 85 86 87 88 89 90 91

Sales Per Square Foot Index

Wal-Mart

Industry

140
130
120
110
100
90

1984 85 86 87 88 89 90 91

Return on Assets (%)

Wal-Mart

Industry

15
14
13
12
11
10
9
8
7
6
5
4

1984 85 86 87 88 89 90 91

Source: The Boston Consulting Group

This strategic vision reached its fullest expression in a largely invisible logistics technique known as "cross-docking." In this system, goods are continuously delivered to Wal-Mart's warehouses, where they are selected, repacked, and then dispatched to stores, often without ever sitting in inventory. Instead of spending valuable time in the warehouse, goods just cross from one loading dock to another in 48 hours or less.

Cross-docking enables Wal-Mart to achieve the economies that come with purchasing full truckloads of goods while avoiding the usual inventory and handling costs. Wal-Mart runs a full 85% of its goods through its warehouse system—as opposed to only 50% for Kmart. This reduces Wal-Mart's costs of sales by 2% to 3% compared with the industry average. That cost difference makes possible the everyday low prices.

But that's not all. Low prices in turn mean that Wal-Mart can save even more by eliminating the expense of frequent promotions. Stable prices also make sales more predictable, thus reducing stockouts and excess inventory. Finally, everyday low prices bring in the customers, which translates into higher sales per retail square foot. These advantages in basic economics make the greeters and the profit sharing easy to afford.

With such obvious benefits, why don't all retailers use cross-docking? The reason: it is extremely difficult to manage. To make cross-docking work, Wal-Mart has had to make strategic investments in a variety of interlocking support systems far beyond what could be justified by conventional ROI criteria.

For example, cross-docking requires continuous contact among Wal-Mart's distribution centers, suppliers, and every point of sale in every store to ensure that orders can flow in and be consolidated and executed within a matter of hours. So Wal-Mart operates a private satellite-communication system that daily sends point-of-sale data directly to Wal-Mart's 4,000 vendors.

Another key component of Wal-Mart's logistics infrastructure is the company's fast and responsive transportation system. The company's 19 distribution centers are serviced by nearly 2,000 company-owned trucks. This dedicated truck fleet permits Wal-Mart to ship goods from warehouse to store in less than 48 hours and to replenish its store shelves twice a week on average. By contrast, the industry norm is once every two weeks.

To gain the full benefits of cross-docking, Wal-Mart has also had to make fundamental changes in its approach to managerial control. Traditionally in the retail industry, decisions about merchandising, pricing, and promotions have been highly centralized and made at the corporate level. Cross-docking, however, turns this command-and-control logic on its head. Instead of the retailer pushing products into the system, customers "pull" products when and where they need them. This approach places a premium on frequent, informal cooperation among stores, distribution centers, and suppliers—with far less centralized control.

The job of senior management at Wal-Mart, then, is not to tell individual store managers what to do but to create an environment where they can learn from the market—and from each other. The company's information systems, for example, provide store managers with detailed information about customer behavior, while a fleet of airplanes regularly ferries store managers to Bentonville, Arkansas headquarters for meetings on market trends and merchandising.

As the company has grown and its stores have multiplied, even Wal-Mart's own private air force hasn't been enough to maintain the necessary contacts among store managers. So Wal-Mart has installed a video link connecting all its stores to corporate headquarters and to each other. Store managers frequently hold videoconferences to exchange information on what's happening in the field, like which products are selling and which ones aren't, which promotions work and which don't.

The final piece of this capabilities mosaic is Wal-Mart's human resources system. The company realizes that its frontline employees play a significant role in satisfying customer needs. So it set out to enhance its organizational capability with programs like stock ownership and profit sharing geared toward making its personnel more responsive to customers. Even the way Wal-Mart stores are organized contributes to this goal. Where Kmart has 5 separate merchandise departments in each store, Wal-Mart has 36. This means that training can be more focused and more effective, and employees can be more attuned to customers.

Kmart did not see its business this way. While Wal-Mart was fine-tuning its business processes and organizational practices, Kmart was following the classic textbook approach that had accounted for its original success. Kmart managed its business by focusing on a few product-centered strategic business units, each a profit center under strong centralized line management. Each SBU made strategy—selecting merchandise, setting prices, and deciding which products to promote. Senior management spent most of its time and resources making line decisions rather than investing in a support infrastructure.

Similarly, Kmart evaluated its competitive advantage at each stage

along a value chain and subcontracted activities that managers concluded others could do better. While Wal-Mart was building its ground
transportation fleet, Kmart was moving *out* of trucking because a
subcontracted fleet was cheaper. While Wal-Mart was building close
relationships with its suppliers, Kmart was constantly switching suppliers in search of price improvements. While Wal-Mart was controlling all the departments in its stores, Kmart was leasing out many of
its departments to other companies on the theory that it could make
more per square foot in rent than through its own efforts.

This is not to say that Kmart managers do not care about their
business processes. After all, they have quality programs too. Nor is it
that Wal-Mart managers ignore the structural dimension of strategy:
they focus on the same consumer segments as Kmart and still have to
make traditional strategic decisions like where to open new stores. The
difference is that Wal-Mart emphasizes behavior—the organizational
practices and business processes in which capabilities are rooted—as
the primary object of strategy and therefore focuses its managerial
attention on the infrastructure that supports capabilities. This subtle
distinction has made all the difference between exceptional and average performance.

Four Principles of Capabilities-Based Competition

The story of Kmart and Wal-Mart illustrates the new paradigm of
competition in the 1990s. In industry after industry, established competitors are being outmaneuvered and overtaken by more dynamic
rivals.

In the years after World War II, Honda was a modest manufacturer of
a 50 cc. engine designed to be attached to a bicycle. Today it is challenging General Motors and Ford for dominance of the global automobile industry.

Xerox invented xerography and the office copier market. But between
1976 and 1982, Canon introduced more than 90 new models, cutting
Xerox's share of the mid-range copier market in half.[1] Today Canon is
a key competitor not only in mid-range copiers but also in high-end
color copiers.

The greatest challenge to department store giants like Macy's comes
neither from other large department stores nor from small boutiques
but from The Limited, a $5.25 billion design, procurement, delivery,

and retailing machine that exploits dozens of consumer segments with the agility of many small boutiques.

Citicorp may still be the largest U.S. bank in terms of assets, but Banc One has consistently enjoyed the highest return on assets in the U.S. banking industry and now enjoys a market capitalization greater than Citicorp's.

These examples represent more than just the triumph of individual companies. They signal a fundamental shift in the logic of competition, a shift that is revolutionizing corporate strategy.

When the economy was relatively static, strategy could afford to be static. In a world characterized by durable products, stable customer needs, well-defined national and regional markets, and clearly identified competitors, competition was a "war of position" in which companies occupied competitive space like squares on a chessboard, building and defending market share in clearly defined product or market segments. The key to competitive advantage was *where* a company chose to compete. *How* it chose to compete was also important but secondary, a matter of execution.

Few managers need reminding of the changes that have made this traditional approach obsolete. As markets fragment and proliferate, "owning" any particular market segment becomes simultaneously more difficult and less valuable. As product life cycles accelerate, dominating existing product segments becomes less important than being able to create new products and exploit them quickly. Meanwhile, as globalization breaks down barriers between national and regional markets, competitors are multiplying and reducing the value of national market share.

In this more dynamic business environment, strategy has to become correspondingly more dynamic. Competition is now a "war of movement" in which success depends on anticipation of market trends and quick response to changing customer needs. Successful competitors move quickly in and out of products, markets, and sometimes even entire businessess—a process more akin to an interactive video game than to chess. In such an environment, the essence of strategy is *not* the structure of a company's products and markets but the dynamics of its behavior. And the goal is to identify and develop the hard-to-imitate organizational capabilities that distinguish a company from its competitors in the eyes of customers.

Companies like Wal-Mart, Honda, Canon, The Limited, or Banc One have learned this lesson. Their experience and that of other successful companies suggest four basic principles of capabilities-based competition:

1. The building blocks of corporate strategy are not products and markets but business processes.
2. Competitive success depends on transforming a company's key processes into strategic capabilities that consistently provide superior value to the customer.
3. Companies create these capabilities by making strategic investments in a support infrastructure that links together and transcends traditional SBUs and functions.
4. Because capabilities necessarily cross functions, the champion of a capabilities-based strategy is the CEO.

A capability is a set of business processes strategically understood. Every company has business processes that deliver value to the customer. But few think of them as the primary object of strategy. Capabilities-based competitors identify their key business processes, manage them centrally, and invest in them heavily, looking for a long-term payback.

Take the example of cross-docking at Wal-Mart. Cross-docking is not the cheapest or the easiest way to run a warehouse. But seen in the broader context of Wal-Mart's inventory-replenishment capability, it is an essential part of the overall process of keeping retail shelves filled while also minimizing inventory and purchasing in truckload quantities.

What transforms a set of indvidual business processes like cross-docking into a strategic capability? The key is to connect them to real customer needs. A capability is strategic only when it begins and ends with the customer.

Of course, just about every company these days claims to be "close to the customer." But there is a qualitative difference in the customer focus of capabilities-driven competitors. These companies conceive of the organization as a giant feedback loop that begins with identifying the needs of the customer and ends with satisfying them.

As managers have grasped the importance of time-based competition, for example, they have increasingly focused on the speed of new product development. But as a unit of analysis, new product *development* is too narrow. It is only part of what is necessary to satisfy a customer and, therefore, to build an organizational capability. Better to think in terms of new product *realization*, a capability that includes the way a product is not only developed but also marketed and serviced. The longer and more complex the string of business processes, the harder it is to transform them into a capability—but the greater the value of that capability once built because competitors have more difficulty imitating it.

Weaving business processes together into organizational capabilities in this way also mandates a new logic of vertical integration. At a time when cost pressures are pushing many companies to outsource more and more activities, capabilities-based competitors are integrating vertically to ensure that they, not a supplier or distributor, control the performance of key business processes. Remember Wal-Mart's decision to own its transportation fleet in contrast to Kmart's decision to subcontract.

Even when a company doesn't actually own every link of the capability chain, the capabilities-based competitor works to tie these parts into its own business systems. Consider Wal-Mart's relationships with its suppliers. In order for Wal-Mart's inventory-replenishment capability to work, vendors have to change their own business processes to be more responsive to the Wal-Mart system. In exchange, they get far better payment terms from Wal-Mart than they do from other discount retailers. At Wal-Mart, the average "days payable," the time between the receipt of an invoice from a supplier and its payment, is 29 days. At Kmart, it is 45.

Another attribute of capabilities is that they are collective and cross-functional—a small part of many people's jobs, not a large part of a few. This helps explain why most companies underexploit capabilities-based competition. Because a capability is "everywhere and nowhere," no one executive controls it entirely. Moreover, leveraging capabilities requires a panoply of strategic investments across SBUs and functions far beyond what traditional cost-benefit metrics can justify. Traditional internal accounting and control systems often miss the strategic nature of such investments. For these reasons, building strategic capabilities cannot be treated as an operating matter and left to operating managers, to corporate staff, or still less to SBU heads. It is the primary agenda of the CEO.

Only the CEO can focus the entire company's attention on creating capabilities that serve customers. Only the CEO can identify and authorize the infrastructure investments on which strategic capabilities depend. Only the CEO can insulate individual managers from any short-term penalties to the P&Ls of their operating units that such investments might bring about.

Indeed, a CEO's success in building and managing capabilities will be the chief test of management skill in the 1990s. The prize will be companies that combine scale and flexibility to outperform the competition along five dimensions:

Speed. The ability to respond quickly to customer or market demands and to incorporate new ideas and technologies quickly into products.

Consistency. The ability to produce a product that unfailingly satisfies customers' expectations.

Acuity. The ability to see the competitive environment clearly and thus to anticipate and respond to customers' evolving needs and wants.

Agility. The ability to adapt simultaneously to many different business environments.

Innovativeness. The ability to generate new ideas and to combine existing elements to create new sources of value.

Becoming a Capabilities-Based Competitor

Few companies are fortunate enough to begin as capabilities-based competitors. For most, the challenge is to become one.

The starting point is for senior managers to undergo the fundamental shift in perception that allows them to see their business in terms of strategic capabilities. Then they can begin to identify and link together essential business processes to serve customer needs. Finally, they can reshape the organization—including managerial roles and responsiblities—to encourage the new kind of behavior necessary to make capabilities-based competition work.

The experience of a medical-equipment company we'll call Medequip illustrates this change process. An established competitor, Medequip recently found itself struggling to regain market share it had lost to a new competitor. The rival had introduced a lower priced, lower performance version of the company's most popular product. Medequip had developed a similar product in response, but senior managers were hesitant to launch it.

Their reasoning made perfect sense according to the traditional competitive logic. As managers saw it, the company faced a classic no-win situation. The new product was lower priced but also lower profit. If the company promoted it aggressively to regain market share, overall profitability would suffer.

But when Medequip managers began to investigate their competitive situation more carefully, they stopped defining the problem in terms of static products and markets. Increasingly, they saw it in terms of the organization's business processes.

Traditionally, the company's functions had operated autonomously. Manufacturing was separate from sales, which was separate from field service. What's more, the company managed field service the way most companies do—as a classic profit center whose resources were deployed to reduce costs and maximize profitability. For instance, Medequip assigned full-time service personnel only to those customers who bought enough equipment to justify the additional cost.

However, a closer look at the company's experience with these steady customers led to a fresh insight: at accounts where Medequip had placed one or more full-time service representatives on-site, the company renewed its highly profitable service contracts at three times the rate of its other accounts. When these accounts needed new equipment, they chose Medequip twice as often as other accounts did and tended to buy the broadest mix of Medequip products as well.

The reason was simple. Medequip's on-site service representatives had become expert in the operations of their customers. They knew what equipment mix best suited the customer and what additional equipment the customer needed. So they had teamed up informally with Medequip's salespeople to become part of the selling process. Because the service reps were on-site full-time, they were also able to respond quickly to equipment problems. And of course, whenever a competitor's equipment broke down, the Medequip reps were on hand to point out the product's shortcomings.

This new knowledge about the dynamics of service delivery inspired top managers to rethink how their company should compete. Specifically, they redefined field service from a stand-alone function to one part of an integrated sales and service capability. They crystallized this new approach in three key business decisions.

First, Medequip decided to use its service personnel *not* to keep costs low but to maximize the life-cycle profitability of a set of targeted accounts. This decision took the form of a dramatic commitment to place at least one service rep on-site with selected customers—no matter how little business each account currently represented.

The decision to guarantee on-site service was expensive, so choosing which customers to target was crucial; there had to be potential for considerable additional business. The company divided its accounts into three categories: those it dominated, those where a single competitor dominated, and those where several competitors were present.

Medequip protected the accounts it dominated by maintaining the already high level of service and by offering attractive terms for renewing service contracts. The company ignored those customers dominated by a single competitor—unless the competitor was having serious problems. All the remaining resources were focused on those accounts where no single competitor had the upper hand.

Next Medequip combined its sales, service, and order-entry organizations into cross-functional teams that concentrated almost exclusively on the needs of the targeted accounts. The company trained service reps in sales techniques so they could take full responsibility for generating new sales leads. This freed up the sales staff to focus on the more strategic role of understanding the long-term needs of the customer's business. Finally, to emphasize Medequip's new commitment to total service, the company even taught its service reps how to fix competitors' equipment.

Once this new organizational structure was in place, Medequip finally introduced its new low price product. The result: the company has not only stopped its decline in market share but also *increased* share by almost 50%. The addition of the lower priced product has reduced profit margins, but the overall mix still includes many higher priced products. And absolute profits are much higher than before.

This story suggests four steps by which any company can transform itself into a capabilities-based competitor:

Shift the strategic framework to achieve aggressive goals. At Medequip, managers transformed what looked like a no-win situation—either lose share or lose profits—into an opportunity for a major competitive victory. They did so by abandoning the company's traditional function, cost, and profit-center orientation and by identifying and managing the capabilities that link customer need to customer satisfaction. The chief expression of this new capabilities-based strategy was the decision to provide on-site service reps to targeted accounts and to create cross-functional sales and service teams.

Organize around the chosen capability and make sure employees have the necessary skills and resources to achieve it. Having set this ambitious competitive goal, Medequip managers next set about reshaping the company in terms of it. Rather than retaining the existing functional structure and trying to encourage coordination through some kind of matrix, they created a brand new organization—Customer Sales and Service—and divided it into "cells" with overall responsibility for specific customers. The company also provided the necessary training so that employees could understand how their new roles would help achieve new business goals. Finally, Medequip created systems to support employees in their new roles. For example, one information system uses CD-ROMs to give field-service personnel quick access to information about Medequip's product line as well as those of competitors.

Make progress visible and bring measurements and reward into alignment. Medequip also made sure that the company's measurement and reward systems reflected the new competitive strategy. Like most companies, the company had never known the profitability of individual customers. Traditionally, field-service employees were measured on overall service profitability. With the shift to the new approach, however, the company had to develop a whole new set of measures—for example, Medequip's "share-by-customer-by-product," the amount of money the company invested in servicing a particular customer, and the customer's current and estimated lifetime profitability. Team members' compensation was calculated according to these new measures.

Do not delegate the leadership of the transformation. Becoming a capabilities-based competitor requires an enormous amount of change. For that reason, it is a process extremely difficult to delegate. Because capabilities are cross-functional, the change process can't be left to middle managers. It requires the hands-on guidance of the CEO and the active involvement of top line managers. At Medequip, the heads of sales, service, and order entry led the subteams that made the actual recommendations, but it was the CEO who oversaw the change process, evaluated their proposals, and made the final decision. His leading role ensured senior management's commitment to the recommended changes.

This top-down change process has the paradoxical result of driving business decision making down to those directly participating in key processes—for example, Medequip's sales and service staff. This leads to a high measure of operational flexibility and an almost reflex-like responsiveness to external change.

A New Logic of Growth: The Capabilities Predator

Once managers reshape the company in terms of its underlying capabilities, they can use these capabilities to define a growth path for the corporation. At the center of capabilities-based competition is a new logic of growth.

In the 1960s, most managers assumed that when growth in a company's basic business slowed, the company should turn to diversification. This was the age of the multibusiness conglomerate. In the 1970s and 1980s, however, it became clear that growth through diversification was difficult. And so, the pendulum of management thinking swung once again. Companies were urged to "stick to their knitting"—that is, to focus on their core business, identify where the profit was, and get rid of everything else. The idea of the corporation became increasingly narrow.

Competing on capabilities provides a way for companies to gain the benefits of both focus and diversification. Put another way, a company that focuses on its strategic capabilities can compete in a remarkable diversity of regions, products, and businesses and do it far more coherently than the typical conglomerate can. Such a company is a "capabilities predator"—able to come out of nowhere and move rapidly from nonparticipant to major player and even to industry leader.

Capabilities-based companies grow by transferring their essential business processes—first to new geographic areas and then to new businesses. Wal-Mart CEO David Glass alludes to this method of growth when he characterizes Wal-Mart as "always pushing from the inside out; we never jump and backfill."

Strategic advantages built on capabilities are easier to transfer geographically than more traditional competitive advantages. Honda, for example, has become a manufacturer in Europe and the United States with relatively few problems. The quality of its cars made in the United States is so good that the company is exporting some of them back to Japan.

In many respects, Wal-Mart's move from small towns in the South to large, northern cities spans as great a cultural gap as Honda's move beyond Japan. And yet, Wal-Mart has done it with barely a hiccup. While the stores are much bigger and the product lines different, the capabilities are exactly the same. Wal-Mart simply replicates its system as soon as the required people are trained. The company estimates that it can train enough new employees to grow about 25% a year.

But the big payoff for capabilities-led growth comes not through geographical expansion but through rapid entry into whole new businesses. Capabilities-based companies do this in at least two ways. The first is by "cloning" their key business processes. Again, Honda is a typical example.

Most people attribute Honda's success to the innovative design of its products or the way the company manufactures them. These factors are certainly important. But the company's growth has been spearheaded by less visible capabilities.

For example, a big part of Honda's original success in motorcycles was due to the company's distinctive capability in "dealer management," which departed from the traditional relationship between motorcycle manufacturers and dealers. Typically, local dealers were motorcycle enthusiasts who were more concerned with finding a way to support their hobby than with building a strong business. They were not particularly interested in marketing, parts-inventory management, or other business systems.

Honda, by contrast, managed its dealers to ensure that they would become successful businesspeople. The company provided operating procedures and policies for merchandising, selling, floor planning, and service management. It trained all its dealers and their entire staffs in these new management systems and supported them with a computerized dealer-management information system. The part-time dealers of competitors were no match for the better prepared and better financed Honda dealers.

Honda's move into new businesses, including lawn mowers, outboard motors, and automobiles, has depended on re-creating this same dealer-management capability in each new sector. Even in segments like luxury cars, where local dealers are generally more service-oriented than those in the motorcycle business, Honda's skill at managing its dealers is transforming service standards. Honda dealers consistently receive the highest ratings for customer satisfaction among auto companies selling in the United States. One reason is that Honda gives its dealers far more autonomy to decide on the spot whether a needed repair is covered by warranty.

How Capabilities Differ from Core Competencies: The Case of Honda

In their influential 1990 HBR article, "The Core Competence of the Corporation," Gary Hamel and C.K. Prahalad mount an attack on traditional notions of strategy that is not so dissimilar from what we are arguing here. For Hamel and Prahalad, however, the central building block of corporate strategy is "core competence." How is a competence different from a capability, and how do the two concepts relate to each other?

Hamel and Prahalad define core competence as the combination of individual technologies and production skills that underlie a company's myriad product lines. Sony's core competence in miniaturization, for example, allows the company to make everything from the Sony Walkman to videocameras to notebook computers. Canon's core competencies in optics, imaging, and microprocessor controls have enabled it to enter markets as seemingly diverse as copiers, laser printers, cameras, and image scanners.

As the above examples suggest, Hamel and Prahalad use core competence to explain the ease with which successful competitors are able to enter new and seemingly unrelated businesses. But a closer look reveals that competencies are not the whole story.

Consider Honda's move from motorcycles into other businesses, including lawn mowers, outboard motors, and automobiles. Hamel and Prahalad attribute Honda's success to its underlying competence in engines and power trains. While Honda's engine competence is certainly important, it alone cannot explain the speed with which the company has successfully moved into a wide range of businesses over the past 20 years. After all, General Motors (to take just one example) is also an accomplished designer and manufacturer of engines. What distinguishes Honda from its competitors is its focus on capabilities.

One important but largely invisible capability is Honda's expertise in "dealer management"—its ability to train and support its dealer network with operating procedures and policies for merchandising, selling, floor planning, and service management. First developed for its motorcycle business, this set of business processes has since been replicated in each new business the company has entered.

Another capability central to Honda's success has been its skill at "product realization." Traditional product development separates planning, proving, and executing into three sequential activities: assessing the market's needs and whether existing products are meeting those needs; testing the proposed product; then building a prototype. The end result of this process is a new factory or organization to introduce the new product. This traditional approach takes a long time—and with time goes money.

Honda has arranged these activities differently. First, planning and proving go on continuously and in parallel. Second, these activities are clearly separated from execution. At Honda, the highly disciplined execution cycle schedules major product revisions every four years and minor revisions every two years. The 1990 Honda Accord, for example, which is the first major redesign of that model since 1986, incorporates a power train developed two years earlier and first used in the 1988 Accord. Finally, when a new product is ready, it is released to *existing* factories and organizations, which dramatically shortens the amount of time needed to launch it. As time is reduced, so are cost and risk.

Consider the following comparison between Honda and GM. In 1984,

Honda launched its Acura division; one year later, GM created Saturn. Honda chose to integrate Acura into its existing organization and facilities. In Europe, for example, the Acura Legend is sold through the same sales force as the Honda Legend. The Acura division now makes three models—the Legend, Integra, and Vigor—and is turning out 300,000 cars a year. At the end of 1991, seven years after it was launched, the division had produced a total of 800,000 vehicles. More important, it had already introduced eight variations of its product line.

By contrast, GM created a separate organization and a separate facility for Saturn. Production began in late 1990, and 1991 will be its first full model year. If GM is lucky, it will be producing 240,000 vehicles in the next year or two and will have two models out.

As the Honda example suggests, competencies and capabilities represent two different but complementary dimensions of an emerging paradigm for corporate strategy. Both concepts emphasize "behavioral" aspects of strategy in contrast to the traditional structural model. But whereas core competence emphasizes technological and production expertise at specific points along the value chain, capabilities are more broadly based, encompassing the entire value chain. In this respect, capabilities are visible to the customer in a way that core competencies rarely are.

Like the "grand unified theory" that modern-day physicists are searching for to explain physical behavior at both the subatomic level and that of the entire cosmos, the combination of core competence and capabilities may define the universal model for corporate strategy in the 1990s and beyond.

But the ultimate form of growth in the capabilities-based company may not be cloning business processes so much as creating processes so flexible and robust that the same set can serve many different businesses. This is the case with Wal-Mart. The company uses the same inventory-replenishment system that makes its discount stores so successful to propel itself into new and traditionally distinct retail sectors.

Take the example of warehouse clubs, no-frills stores that sell products in bulk at a deep discount. In 1983, Wal-Mart created Sam's Club to compete with industry founder Price Club and Kmart's own PACE Membership Warehouse. Within four years, Sam's Club sales had passed those of both Price and PACE, making it the largest wholesale club in the country (see Exhibit II). Sam's 1990 sales were $5.3 billion, compared with $4.9 billion for Price and $1.6 billion for PACE. What's

Exhibit II.

Portrait of a Capabilities Predator

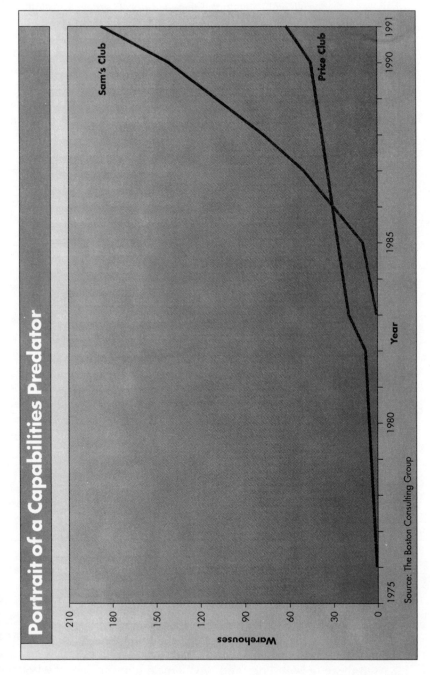

Source: The Boston Consulting Group

more, Wal-Mart has repeated this rapid penetration strategy in other retail sectors, including pharmacies, European-style hypermarkets, and large, no-frills grocery stores known as superstores.

While Wal-Mart has been growing by quickly entering these new businesses, Kmart has tried to grow by acquisition, with mixed success. In the past decade, Kmart has bought and sold a number of companies in unrelated businesses such as restaurants and insurance—an indication the company has had difficulty adding value.

This is not to suggest that growth by acquisition is necessarily doomed to failure. Indeed, the company that is focused on its capabilities is often better able to target sensible acquisitions and then integrate them successfully. For example, Wal-Mart has recently begun to supplement its growth "from the inside out" by acquiring companies—for example, other small warehouse clubs and a retail and grocery distributor—whose operations can be folded into the Wal-Mart system.

It is interesting to speculate where Wal-Mart will strike next. The company's inventory-replenishment capability could prove to be a strong competitive advantage in a wide variety of retail businesses. In the past decade, Wal-Mart came out of nowhere to challenge Kmart. In the next decade, companies such as Toys "R" Us (Wal-Mart already controls as much as 10% of the $13 billion toy market) and Circuit City (consumer electronics) may find themselves in the sights of this capabilities predator.

The Future of Capabilities-Based Competition

For the moment, capabilities-based companies have the advantage of competing against rivals still locked into the old way of seeing the competitive environment. But such a situation won't last forever. As more and more companies make the transition to capabilities-based competition, the simple fact of competing on capabilities will become less important than the specific capabilities a company has chosen to build. Given the necessary long-term investments, the strategic choices managers make will end up determining a company's fate.

If Wal-Mart and Kmart are a good example of the present state of capabilities-based competition, the story of two fast-growing regional banks suggests its future. Wachovia Corporation, with dual headquarters in Winston-Salem, North Carolina and Atlanta, Georgia, has superior returns and growing market share throughout its core markets in both states. Banc One, based in Columbus, Ohio, has consistently enjoyed the highest return on assets in the U.S. banking industry. Both banks compete on capabilities, but they do it in very different ways.

Wachovia competes on its ability to understand and serve the needs of individual customers, a skill that manifests itself in probably the highest "cross-sell ratio"—the average number of products per customer—of any bank in the country. The linchpin of this capability is the company's roughly 600 "personal bankers," frontline employees who provide Wachovia's mass-market customers with a degree of personalized service approaching what has traditionally been available only to private banking clients. The company's specialized support systems allow each personal banker to serve about 1,200 customers. Among those systems: an integrated customer-information file, simplified work processes that allow the bank to respond to almost all customer requests by the end of business that day, and a five-year personal banker training program.

Where Wachovia focuses on meeting the needs of individual customers, Banc One's distinctive ability is to understand and respond to the needs of entire *communities*. To do community banking effectively, a bank has to have deep roots in the local community. But traditionally, local banks have not been able to muster the professional expertise, state-of-the-art products, and highly competitive cost structure of large national banks like Citicorp. Banc One competes by offering its customers the best of both these worlds. Or in the words of one company slogan, Banc One "out-locals the national banks and out-nationals the local banks."

Striking this balance depends on two factors. One is local autonomy. The central organizational role in the Banc One business system is played not by frontline employees but by the presidents of the 51 affiliate banks in the Banc One network. Affiliate presidents have exceptional power within their own region. They select products, establish prices and marketing strategy, make credit decisions, and set internal management policies. They can even overrule the activities of Banc One's centralized direct-marketing businesses. But while Banc One's affiliate system is highly decentralized, its success also depends on an elaborate, and highly centralized, process of continuous organizational learning. Affiliate presidents have the authority to mold bank products and services to local conditions, but they are also expected to learn from best practice throughout the Banc One system and to adapt it to their own operations.

Banc One collects an extraordinary amount of detailed and current information on each affiliate bank's internal and external perform-ance. For example, the bank regularly publishes "league tables" on numerous measures of operating performance, with the worst per-formers listed first. This encourages collaboration to improve the weakest affiliates rather than competition to be the best. The bank also continuously engages in workflow re-engineering and process sim-plification. The 100 most successful projects, known as the "Best of the Best," are documented and circulated among affiliates.

Wachovia and Banc One both compete on capabilities. Both banks focus on key business processes and place critical decision-making authority with the people directly responsible for them. Both manage these processes through a support system that spans the traditional functional structure, and senior managers concentrate on managing this system rather than controlling decisions. Both are decentralized but focused, single-minded but flexible.

But there the similarities end. Wachovia responds to individual customers en masse with personalization akin to that of a private banker. Banc One responds to local markets en masse with the flexibil-ity and canniness of the traditional community bank. As a result, they focus on different business processes: Wachovia on the transfer of customer-specific information across numerous points of customer contact; Banc One on the transfer of best practices across affiliate banks. They also empower different levels in the organization: the personal banker at Wachovia, the affiliate president at Banc One.

Most important, they grow differently. Because so much of Wacho-via's capability is embedded in the training of the personal bankers, the bank has made few acquisitions and can integrate them only very slowly. Banc One's capabilities, by contrast, are especially easy to transfer to new acquisitions. All the company needs to do is install its corporate MIS and intensively train the acquired bank's senior officers, a process that can be done in a few months, as opposed to the much longer period it takes Wachovia to train a new cadre of frontline bankers. Banc One has therefore made acquisitions almost a separate line of business.

If Banc One and Wachovia were to compete against each other, it is not clear who would win. Each would have strengths that the other could not match. Wachovia's capability to serve individual customers by cross-selling a wide range of banking products will in the long term probably allow the company to extract more profit per customer than Banc One. On the other hand, Wachovia cannot adapt its products, pricing, and promotion to local market conditions the way Banc One

3
The Knowledge-Creating Company

Ikujiro Nonaka

In an economy where the only certainty is uncertainty, the one sure source of lasting competitive advantage is knowledge. When markets shift, technologies proliferate, competitors multiply, and products become obsolete almost overnight, successful companies are those that consistently create new knowledge, disseminate it widely throughout the organization, and quickly embody it in new technologies and products. These activities define the "knowledge-creating" company, whose sole business is continuous innovation.

And yet, despite all the talk about "brainpower" and "intellectual capital," few managers grasp the true nature of the knowledge-creating company—let alone know how to manage it. The reason: they misunderstand what knowledge is and what companies must do to exploit it.

Deeply ingrained in the traditions of Western management, from Frederick Taylor to Herbert Simon, is a view of the organization as a machine for "information processing." According to this view, the only useful knowledge is formal and systematic—hard (read: quantifiable) data, codified procedures, universal principles. And the key metrics for measuring the value of new knowledge are similarly hard and quantifiable—increased efficiency, lower costs, improved return on investment.

But there is another way to think about knowledge and its role in business organizations. It is found most commonly at highly successful Japanese competitors like Honda, Canon, Matsushita, NEC, Sharp, and Kao. These companies have become famous for their ability to respond quickly to customers, create new markets, rapidly develop new products, and dominate emergent technologies. The secret of their success is their unique approach to managing the creation of new knowledge.

To Western managers, the Japanese approach often seems odd or even incomprehensible. Consider the following examples:

> How is the slogan "Theory of Automobile Evolution" a meaningful design concept for a new car? And yet, this phrase led to the creation of the Honda City, Honda's innovative urban car.
>
> Why is a beer can a useful analogy for a personal copier? Just such an analogy caused a fundamental breakthrough in the design of Canon's revolutionary mini-copier, a product that created the personal copier market and has led Canon's successful migration from its stagnating camera business to the more lucrative field of office automation.
>
> What possible concrete sense of direction can a made-up word such as "optoelectronics" provide a company's product-development engineers? Under this rubric, however, Sharp has developed a reputation for creating "first products" that define new technologies and markets, making Sharp a major player in businesses ranging from color televisions to liquid crystal displays to customized integrated circuits.

In each of these cases, cryptic slogans that to a Western manager sound just plain silly—appropriate for an advertising campaign perhaps but certainly not for running a company—are in fact highly effective tools for creating new knowledge. Managers everywhere recognize the serendipitous quality of innovation. Executives at these Japanese companies are *managing* that serendipity to the benefit of the company, its employees, and its customers.

The centerpiece of the Japanese approach is the recognition that creating new knowledge is not simply a matter of "processing" objective information. Rather, it depends on tapping the tacit and often highly subjective insights, intuitions, and hunches of individual employees and making those insights available for testing and use by the company as a whole. The key to this process is personal commitment, the employees' sense of identity with the enterprise and its mission. Mobilizing that commitment and embodying tacit knowledge in actual technologies and products require managers who are as comfortable with images and symbols—slogans such as Theory of Automobile Evolution, analogies like that between a personal copier and a beer can, metaphors such as "optoelectronics"—as they are with hard numbers measuring market share, productivity, or ROI.

The more holistic approach to knowledge at many Japanese companies is also founded on another fundamental insight. A company is not a machine but a living organism. Much like an individual, it can have a collective sense of identity and fundamental purpose. This is the organizational equivalent of self-knowledge—a shared understanding of what the company stands for, where it is going, what kind of world it wants to live in, and, most important, how to make that world a reality.

In this respect, the knowledge-creating company is as much about ideals as it is about ideas. And that fact fuels innovation. The essence of innovation is to re-create the world according to a particular vision or ideal. To create new knowledge means quite literally to re-create the company and everyone in it in a nonstop process of personal and organizational self-renewal. In the knowledge-creating company, inventing new knowledge is not a specialized activity—the province of the R&D department or marketing or strategic planning. It is a way of behaving, indeed a way of being, in which everyone is a knowledge worker—that is to say, an entrepreneur.

The reasons why Japanese companies seem especially good at this kind of continuous innovation and self-renewal are complicated. But the key lesson for managers is quite simple: much as manufacturers around the world have learned from Japanese manufacturing techniques, any company that wants to compete on knowledge must also learn from Japanese techniques of knowledge creation. The experiences of the Japanese companies discussed below suggest a fresh way to think about managerial roles and responsibilities, organizational design, and business practices in the knowledge-creating company. It is an approach that puts knowledge creation exactly where it belongs: at the very center of a company's human resources strategy.

The Spiral of Knowledge

New knowledge always begins with the individual. A brilliant researcher has an insight that leads to a new patent. A middle manager's intuitive sense of market trends becomes the catalyst for an important new product concept. A shop-floor worker draws on years of experience to come up with a new process innovation. In each case, an individual's personal knowledge is transformed into organizational knowledge valuable to the company as a whole.

Making personal knowledge available to others is the central activity of the knowledge-creating company. It takes place continuously and at all levels of the organization. And as the following example suggests, sometimes it can take unexpected forms.

In 1985, product developers at the Osaka-based Matsushita Electric

Company were hard at work on a new home bread-making machine. But they were having trouble getting the machine to knead dough correctly. Despite their efforts, the crust of the bread was overcooked while the inside was hardly done at all. Employees exhaustively analyzed the problem. They even compared X rays of dough kneaded by the machine and dough kneaded by professional bakers. But they were unable to obtain any meaningful data.

Finally, software developer Ikuko Tanaka proposed a creative solution. The Osaka International Hotel had a reputation for making the best bread in Osaka. Why not use it as a model? Tanaka trained with the hotel's head baker to study his kneading technique. She observed that the baker had a distinctive way of stretching the dough. After a year of trial and error, working closely with the project's engineers, Tanaka came up with product specifications—including the addition of special ribs inside the machine—that successfully reproduced the baker's stretching technique and the quality of the bread she had learned to make at the hotel. The result: Matsushita's unique "twist dough" method and a product that in its first year set a record for sales of a new kitchen appliance.

Ikuko Tanaka's innovation illustrates a movement between two very different types of knowledge. The end point of that movement is "explicit" knowledge: the product specifications for the bread-making machine. Explicit knowledge is formal and systematic. For this reason, it can be easily communicated and shared, in product specifications or a scientific formula or a computer program.

But the starting point of Tanaka's innovation is another kind of knowledge that is not so easily expressible: "tacit" knowledge like that possessed by the chief baker at the Osaka International Hotel. Tacit knowledge is highly personal. It is hard to formalize and, therefore, difficult to communicate to others. Or in the words of the philosopher Michael Polanyi, "We can know more than we can tell." Tacit knowledge is also deeply rooted in action and in an individual's commitment to a specific context—a craft or profession, a particular technology or product market, or the activities of a work group or team.

Tacit knowledge consists partly of technical skills—the kind of informal, hard-to-pin-down skills captured in the term "know-how." A master craftsman after years of experience develops a wealth of expertise "at his fingertips." But he is often unable to articulate the scientific or technical principles behind what he knows.

At the same time, tacit knowledge has an important cognitive dimension. It consists of mental models, beliefs, and perspectives so ingrained that we take them for granted, and therefore cannot easily articulate them. For this very reason, these implicit models profoundly shape how we perceive the world around us.

The distinction between tacit and explicit knowledge suggests four basic patterns for creating knowledge in any organization:

1. From Tacit to Tacit. Sometimes, one individual shares tacit knowledge directly with another. For example, when Ikuko Tanaka apprentices herself to the head baker at the Osaka International Hotel, she learns his tacit skills through observation, imitation, and practice. They become part of her own tacit knowledge base. Put another way, she is "socialized" into the craft.

But on its own, socialization is a rather limited form of knowledge creation. True, the apprentice learns the master's skills. But neither the apprentice nor the master gain any systematic insight into their craft knowledge. Because their knowledge never becomes explicit, it cannot easily be leveraged by the organization as a whole.

2. From Explicit to Explicit. An individual can also combine discrete pieces of explicit knowledge into a new whole. For example, when a comptroller of a company collects information from throughout the organization and puts it together in a financial report, that report is new knowledge in the sense that it synthesizes information from many different sources. But this combination does not really extend the company's existing knowledge base either.

But when tacit and explicit knowledge interact, as in the Matsushita example, something powerful happens. It is precisely this exchange *between* tacit and explicit knowledge that Japanese companies are especially good at developing.

3. From Tacit to Explicit. When Ikuko Tanaka is able to articulate the foundations of her tacit knowledge of bread making, she converts it into explicit knowledge, thus allowing it to be shared with her project-development team. Another example might be the comptroller who, instead of merely compiling a conventional financial plan for his company, develops an innovative new approach to budgetary control based on his own tacit knowledge developed over years in the job.

4. From Explicit to Tacit. What's more, as new explicit knowledge is shared throughout an organization, other employees begin to internalize it—that is, they use it to broaden, extend, and reframe their own tacit knowledge. The comptroller's proposal causes a revision of the company's financial control system. Other employees use the innovation and eventually come to take it for granted as part of the background of tools and resources necessary to do their jobs.

In the knowledge-creating company, all four of these patterns exist in dynamic interaction, a kind of spiral of knowledge. Think back to Matsushita's Ikuko Tanaka:

1. First, she learns the tacit secrets of the Osaka International Hotel baker (socialization).
2. Next, she translates these secrets into explicit knowledge that she can communicate to her team members and others at Matsushita (articulation).
3. The team then standardizes this knowledge, putting it together into a manual or workbook and embodying it in a product (combination).
4. Finally, through the experience of creating a new product, Tanaka and her team members enrich their own tacit knowledge base (internalization). In particular, they come to understand in an extremely intuitive way that products like the home bread-making machine can provide genuine quality. That is, the machine must make bread that is as good as that of a professional baker.

This starts the spiral of knowledge all over again, but this time at a higher level. The new tacit insight about genuine quality developed in designing the home bread-making machine is informally conveyed to other Matsushita employees. They use it to formulate equivalent quality standards for other new Matsushita products—whether kitchen appliances, audiovisual equipment, or white goods. In this way, the organization's knowledge base grows ever broader.

Articulation (converting tacit knowledge into explicit knowledge) and internalization (using that explicit knowledge to extend one's own tacit knowledge base) are the critical steps in this spiral of knowledge. The reason is that both require the active involvement of the self—that is, personal commitment. Ikuko Tanaka's decision to apprentice herself to a master baker is one example of this commitment. Similarly, when the comptroller articulates his tacit knowledge and embodies it in a new innovation, his personal identity is directly involved in a way it is not when he merely "crunches" the numbers of a conventional financial plan.

Indeed, because tacit knowledge includes mental models and beliefs in addition to know-how, moving from the tacit to the explicit is really a process of articulating one's vision of the world—what it is and what it ought to be. When employees invent new knowledge, they are also reinventing themselves, the company, and even the world.

When managers grasp this, they realize that the appropriate tools for managing the knowledge-creating company look very different from those found at most Western companies.

From Metaphor to Model

To convert tacit knowledge into explicit knowledge means finding a way to express the inexpressible. Unfortunately, one of the most powerful management tools for doing so is also among the most frequently overlooked: the store of figurative language and symbolism that managers can draw from to articulate their intuitions and insights. At Japanese companies, this evocative and sometimes extremely poetic language figures especially prominently in product development.

In 1978, top management at Honda inaugurated the development of a new-concept car with the slogan, "Let's gamble." The phrase expressed senior executives' conviction that Honda's Civic and the Accord models were becoming too familiar. Managers also realized that along with a new postwar generation entering the car market, a new generation of young product designers was coming of age with unconventional ideas about what made a good car.

The business decision that followed from the "Let's gamble" slogan was to form a new-product development team of young engineers and designers (the average age was 27). Top management charged the team with two—and only two—instructions: first, to come up with a product concept fundamentally different from anything the company had ever done before; and second, to make a car that was inexpensive but not cheap.

This mission might sound vague, but in fact it provided the team an extremely clear sense of direction. For instance, in the early days of the project, some team members proposed designing a smaller and cheaper version of the Honda Civic—a safe and technologically feasible option. But the team quickly decided this approach contradicted the entire rationale of its mission. The only alternative was to invent something totally new.

Project team leader Hiroo Watanabe coined another slogan to express his sense of the team's ambitious challenge: Theory of Automobile Evolution. The phrase described an ideal. In effect, it posed the question: If the automobile were an organism, how should it evolve? As team members argued and discussed what Watanabe's slogan might possibly mean, they came up with an answer in the form of yet another slogan: "man-maximum, machine-minimum." This captured the team's belief that the ideal car should somehow transcend the traditional human-machine relationship. But that required challenging what Watanabe called "the reasoning of Detroit," which had sacrificed comfort for appearance.

The "evolutionary" trend the team articulated eventually came to be embodied in the image of a sphere—a car simultaneously "short" (in length) and "tall" (in height). Such a car, they reasoned, would be lighter and cheaper, but also more comfortable and more solid than traditional cars. A sphere provided the most room for the passenger while taking up the least amount of space on the road. What's more, the shape minimized the space taken up by the engine and other mechanical systems. This gave birth to a product concept the team called "Tall Boy," which eventually led to the Honda City, the company's distinctive urban car.

The Tall Boy concept totally contradicted the conventional wisdom about automobile design at the time, which emphasized long, low sedans. But the City's revolutionary styling and engineering were prophetic. The car inaugurated a whole new approach to design in the Japanese auto industry based on the man-maximum, machine-minimum concept, which has led to the new generation of "tall and short" cars now quite prevalent in Japan.

The story of the Honda City suggests how Japanese companies use figurative language at all levels of the company and in all phases of the product development process. It also begins to suggest the different kinds of figurative language and the distinctive role each plays.

One kind of figurative language that is especially important is metaphor. By "metaphor," I don't just mean a grammatical structure or allegorical expression. Rather, metaphor is a distinctive method of perception. It is a way for individuals grounded in different contexts and with different experiences to understand something intuitively through the use of imagination and symbols without the need for analysis or generalization. Through metaphors, people put together what they know in new ways and begin to express what they know but cannot yet say. As such, metaphor is highly effective in fostering direct commitment to the creative process in the early stages of knowledge creation.

Metaphor accomplishes this by merging two different and distant areas of experience into a single, inclusive image or symbol—what linguistic philosopher Max Black has aptly described as "two ideas in one phrase." By establishing a connection between two things that seem only distantly related, metaphors set up a discrepancy or conflict.

Often, metaphoric images have multiple meanings, appear logically contradictory or even irrational. But far from being a weakness, this is in fact an enormous strength. For it is the very conflict that metaphors embody that jump-starts the creative process. As employees try to define more clearly the insight that the metaphor expresses, they work to reconcile the conflicting meanings. That is the first step in making the tacit explicit.

Consider the example of Hiroo Watanabe's slogan, Theory of Automobile Evolution. Like any good metaphor, it combines two ideas one wouldn't normally think of together—the automobile, which is a machine, and the theory of evolution, which refers to living organisms. And yet, this discrepancy is a fruitful platform for speculation about the characteristics of the ideal car.

But while metaphor triggers the knowledge-creation process, it alone is not enough to complete it. The next step is analogy. Whereas metaphor is mostly driven by intuition and links images that at first glance seem remote from each other, analogy is a more structured process of reconciling contradictions and making distinctions. Put another way, by clarifying how the two ideas in one phrase actually are alike and not alike, the contradictions incorporated into metaphors are harmonized by analogy. In this respect, analogy is an intermediate step between pure imagination and logical thinking.

Probably the best example of analogy comes from the development of Canon's revolutionary mini-copier. Canon designers knew that for the first personal copier to be successful, it had to be reliable. To ensure reliability, they proposed to make the product's photosensitive copier drum—which is the source of 90% of all maintenance problems—disposable. To be disposable, however, the drum would have to be easy and cheap to make. How to manufacture a throwaway drum?

The breakthrough came one day when task-force leader Hiroshi Tanaka ordered out for some beer. As the team discussed design problems over their drinks, Tanaka held one of the beer cans and wondered aloud, "How much does it cost to manufacture this can?" The question led the team to speculate whether the same process for making an aluminum beer can could be applied to the manufacture of an aluminum copier drum. By exploring how the drum actually is and is not like a beer can, the mini-copier development team was able to come up with the process technology that could manufacture an aluminum copier drum at the appropriate low cost.

Finally, the last step in the knowledge-creation process is to create an actual model. A model is far more immediately conceivable than a metaphor or an analogy. In the model, contradictions get resolved and concepts become transferable through consistent and systematic logic. The quality standards for the bread at the Osaka International Hotel lead Matsushita to develop the right product specifications for its home bread-making machine. The image of a sphere leads Honda to its Tall Boy product concept.

Of course, terms like "metaphor," "analogy," and "model" are ideal types. In reality, they are often hard to distinguish from each other; the same phrase or image can embody more than one of the three functions. Still, the three terms capture the process by which organizations convert tacit knowledge into explicit knowledge: first, by linking contradictory things and ideas through metaphor; then, by resolving these contradictions through analogy; and, finally, by crystallizing the created concepts and embodying them in a model, which makes the knowledge available to the rest of the company.

From Chaos to Concept: Managing the Knowledge-Creating Company

Understanding knowledge creation as a process of making tacit knowledge explicit—a matter of metaphors, analogies, and models—has direct implications for how a company designs its organization and defines managerial roles and responsibilities within it. This is the "how" of the knowledge-creating company, the structures and practices that translate a company's vision into innovative technologies and products.

The fundamental principle of organizational design at the Japanese companies I have studied is redundancy—the conscious overlapping of company information, business activities, and managerial responsibilities. To Western managers, the term "redundancy," with its connotations of unnecessary duplication and waste, may sound unappealing. And yet, building a redundant organization is the first step in managing the knowledge-creating company.

Redundancy is important because it encourages frequent dialogue and communication. This helps create a "common cognitive ground" among employees and thus facilitates the transfer of tacit knowledge. Since members of the organization share overlapping information, they can sense what others are struggling to articulate. Redundancy also spreads new explicit knowledge through the organization so it can be internalized by employees.

The organizational logic of redundancy helps explain why Japanese companies manage product development as an overlapping process where different functional divisions work together in a shared division of labor. At Canon, redundant product development goes one step further. The company organizes product-development teams according to "the principle of internal competition." A team is divided into competing groups that develop different approaches to the same project and then argue over the advantages and disadvantages of their proposals. This encourages the team to look at a project from a variety of perspectives. Under the guidance of a team leader, the team eventually develops a common understanding of the "best" approach.

In one sense, such internal competition is wasteful. Why have two or more groups of employees pursuing the same product-development project? But when responsibilities are shared, information proliferates, and the organization's ability to create and implement concepts is accelerated.

At Canon, for example, inventing the mini-copier's low-cost disposable drum resulted in new technologies that facilitated miniaturization, weight reduction, and automated assembly. These technologies were then quickly applied to other office automation products such as microfilm readers, laser printers, word processors, and typewriters. This was an important factor in diversifying Canon from cameras to office automation and in securing a competitive edge in the laser printer industry. By 1987—only five years after the mini-copier was introduced—a full 74% of Canon's revenues came from its business machines division.

Another way to build redundancy is through strategic rotation, especially between different areas of technology and between functions such as R&D and marketing. Rotation helps employees understand the business from a multiplicity of perspectives. This makes organizational knowledge more "fluid" and easier to put into practice. At Kao Corporation, a leading Japanese consumer-products manufacturer, researchers often "retire" from the R&D department by the age of 40 in order to transfer to other departments such as marketing, sales, or production. And all employees are expected to hold at least three different jobs in any given ten-year period.

Free access to company information also helps build redundancy. When information differentials exist, members of an organization can no longer interact on equal terms, which hinders the search for different interpretations of new knowledge. Thus Kao's top management does not allow any discrimination in access to information among employees. All company information (with the exception of personnel data) is stored in a single integrated database, open to any employee regardless of position.

As these examples suggest, no one department or group of experts has the exclusive responsibility for creating new knowledge in the knowledge-creating company. Senior managers, middle managers, and frontline employees all play a part. Indeed, the value of any one person's contribution is determined less by his or her location in the organizational hierarchy than by the importance of the information he or she provides to the entire knowledge-creating system.

But this is not to say that there is no differentiation among roles and responsibilities in the knowledge-creating company. In fact, creating new knowledge is the product of a dynamic interaction among three roles.

Frontline employees are immersed in the day-to-day details of particular technologies, products, or markets. No one is more expert in the realities of a company's business than they are. But while these employees are deluged with highly specific information, they often find it extremely difficult to turn that information into useful knowledge. For one thing, signals from the marketplace can be vague and ambiguous. For another, employees can become so caught up in their own narrow perspective, that they lose sight of the broader context.

What's more, even when employees *do* develop meaningful ideas and insights, it can still be difficult to communicate the import of that information to others. People don't just passively receive new knowledge, they actively interpret it to fit their own situation and perspective. Thus what makes sense in one context can change or even lose its meaning when communicated to people in a different context. As a result, there is a continual shift in meaning as new knowledge is diffused in an organization.

The confusion created by the inevitable discrepancies in meaning that occur in any organization might seem like a problem. In fact, it can be a rich source of new knowledge—*if* a company knows how to manage it. The key to doing so is continuously challenging employees to reexamine what they take for granted. Such reflection is always necessary in the knowledge-creating company, but it is especially essential during times of crisis or breakdown, when a company's traditional categories of knowledge no longer work. At such moments, ambiguity can prove extremely useful as a source of alternative meanings, a fresh way to think about things, a new sense of direction. In this respect, new knowledge is born in chaos.

The main job of managers in the knowledge-creating company is to orient this chaos toward purposeful knowledge creation. Managers do this by providing employees with a conceptual framework that helps them make sense of their own experience. This takes place at the senior management level at the top of the company and at the middle management level on company teams.

Senior managers give voice to a company's future by articulating metaphors, symbols, and concepts that orient the knowledge-creating activities of employees. They do this by asking the questions: What are we trying to learn? What do we need to know? Where should we be going? Who are we? If the job of frontline employees is to know "what is," then the job of senior executives is to know "what ought to be." Or in the words of Hiroshi Honma, senior researcher at Honda: "Senior managers are romantics who go in quest of the ideal."

At some of the Japanese companies I have studied, CEOs talk about this role in terms of their responsibility for articulating the company's "conceptual umbrella": the grand concepts that in highly universal and abstract terms identify the common features linking seemingly disparate activities or businesses into a coherent whole. Sharp's dedication to optoelectronics is a good example.

In 1973, Sharp invented the first low-power electronic calculator by combining two key technologies—liquid crystal displays (LCDs) and complementary metal oxide semiconductors (CMOSs). Company technologists coined the term "optoelectronics" to describe this merging of microelectronics with optical technologies. The company's senior managers then took up the word and magnified its impact far beyond the R&D and engineering departments in the company.

Optoelectronics represents an image of the world that Sharp wants to live in. It is one of the key concepts articulating what the company ought to be. As such, it has become an overarching guide for the company's strategic development. Under this rubric, Sharp has moved beyond its original success in calculators to become a market leader in a broad range of products based on LCD and semiconductor technologies, including: the Electronic Organizer pocket notebook, LCD projection systems, as well as customized integrated circuits such as masked ROMs, ASICs, and CCDs (charge-coupled devices, which convert light into electronic signals).

Other Japanese companies have similar umbrella concepts. At NEC, top management has categorized the company's knowledge base in terms of a few key technologies and then developed the metaphor "C&C" (for "computers and communications"). At Kao, the umbrella concept is "surface active science," referring to techniques for coating the surface area of materials. This phrase has guided the company's diversification into products ranging from soap detergents to cosmetics to floppy disks—all natural derivatives of Kao's core knowledge base.

Another way top management provides employees with a sense of direction is by setting the standards for justifying the value of the knowledge that is constantly being developed by the organization's members. Deciding which efforts to support and develop is a highly strategic task.

In most companies, the ultimate test for measuring the value of new knowledge is economic—increased efficiency, lower costs, improved ROI. But in the knowledge-creating company, other more qualitative factors are equally important. Does the idea embody the company's vision? Is it an expression of top management's aspirations and strategic goals? Does it have the potential to build the company's organizational knowledge network?

The decision by Mazda to pursue the development of the rotary engine is a classic example of this more qualitative kind of justification. In 1974, the product-development team working on the engine was facing heavy pressure within the company to abandon the project. The rotary engine was a "gas guzzler," critics complained. It would never succeed in the marketplace.

Kenichi Yamamoto, head of the development team (and currently Mazda's chairman), argued that to stop the project would mean giving up on the company's dream of revolutionizing the combustion engine. "Let's think this way," Yamamoto proposed. "We are making history, and it is our fate to deal with this challenge." The decision to continue led to Mazda's successful rotary-engine sports car, the Savanna RX-7.

Seen from the perspective of traditional management, Yamamoto's argument about the company's "fate" sounds crazy. But in the context of the knowledge-creating company, it makes perfect sense. Yamamoto appealed to the fundamental aspirations of the company— what he termed "dedication to uncompromised value"—and to the strategy of technological leadership that senior executives had articulated. He showed how the rotary-engine project enacted the organization's commitment to its vision. Similarly, continuing the project reinforced the individual commitment of team members to that vision and to the organization.

Umbrella concepts and qualitative criteria for justification are crucial to giving a company's knowledge-creating activities a sense of direction. And yet, it is important to emphasize that a company's vision needs also to be open-ended, susceptible to a variety of different and even conflicting interpretations. At first glance, this may seem contradictory. After all, shouldn't a company's vision be unambiguous, coherent, and clear? If a vision is *too* unambiguous, however, it becomes more akin to an order or an instruction. And orders do not foster the high degree of personal commitment on which effective knowledge creation depends.

A more equivocal vision gives employees and work groups the freedom and autonomy to set their own goals. This is important because while the ideals of senior management are important, on their own they are not enough. The best that top management can do is to clear away any obstacles and prepare the ground for self-organizing groups or teams. Then, it is up to the teams to figure out what the ideals of the top mean in reality. Thus at Honda, a slogan as vague as "Let's gamble" and an extremely broad mission gave the Honda City product-development team a strong sense of its own identity, which led to a revolutionary new product.

Teams play a central role in the knowledge-creating company because they provide a shared context where individuals can interact with each other and engage in the constant dialogue on which effective reflection depends. Team members create new points of view through dialogue and discussion. They pool their information and examine it from various angles. Eventually, they integrate their diverse individual perspectives into a new collective perspective.

This dialogue can—indeed, should—involve considerable conflict and disagreement. It is precisely such conflict that pushes employees to question existing premises and make sense of their experience in a new way. "When people's rhythms are out of sync, quarrels occur and it's hard to bring people together," acknowledges a deputy manager for advanced technology development at Canon. "Yet if a group's rhythms are completely in unison from the beginning, it's also difficult to achieve good results."

As team leaders, middle managers are at the intersection of the vertical and horizontal flows of information in the company. They serve as a bridge between the visionary ideals of the top and the often chaotic market reality of those on the front line of the business. By creating middle-level business and product concepts, middle managers mediate between "what is" and "what should be." They remake reality according to the company's vision.

Thus at Honda, top management's decision to try something completely new took concrete form at the level of Hiroo Watanabe's product-development team in the Tall Boy product concept. At Canon, the company aspiration, "Making an excellent company through transcending the camera business," became a reality when Hiroshi Tanaka's task force developed the "Easy Maintenance" product concept, which eventually gave birth to the personal copier. And at Matsushita, the company's grand concept, "Human Electronics," came to life through the efforts of Ikuko Tanaka and others who developed the middle-range concept, "Easy Rich," and embodied it in the automatic bread-making machine.

In each of these cases, middle managers synthesized the tacit knowledge of both frontline employees and senior executives, made it explicit, and incorporated it into new technologies and products. In this respect, they are the true "knowledge engineers" of the knowledge-creating company.

4

Message and Muscle: An Interview with Swatch Titan Nicolas Hayek

William Taylor

Nicolas G. Hayek is a rare phenomenon in Europe—a genuine business celebrity. In his home country of Switzerland, and increasingly across the Continent, he is an engaging presence in newspapers and magazines and on television talk shows. Hayek, 64, has earned his fame. He and his colleagues at the Swiss Corporation for Microelectronics and Watchmaking (SMH) have engineered one of the most spectacular industrial comebacks in the world—the revitalization of the Swiss watch industry.

The dimensions of the turnaround are staggering. SMH took shape in 1983 when Hayek recommended that Switzerland's banks merge the country's two giant (and insolvent) watch manufacturers. That year, the new company generated revenues of SFr1.5 billion ($1.1 billion) and lost SFr173 million ($124 million). In 1992, SMH generated revenue of about SFr3 billion ($2.1 billion) and posted profits of more than SFr400 million ($286 million).

Nicolas Hayek has been involved with SMH from the beginning. In the early 1980s, the banks named Hayek their chief adviser on the troubled watch industry. He was already well known as the founder and CEO of Hayek Engineering, a Zurich-based consulting firm that is something of a cross between Arthur D. Little and McKinsey & Company. In 1985, the banks proposed that Hayek buy a controlling equity stake in SMH. He assembled a group of investors, retained the single largest stake, and became CEO.

The original investors paid SFr100 per share. Today SMH trades at SFr1,500 per share. Its market value exceeds SFr4.9 billion ($3.5

billion). Hayek's personal stake is worth roughly SFr1 billion ($700 million).

But SMH is more than a turnaround story. It is a case study of Hayek's management philosophy and strategic thinking, both of which are strikingly at odds with the prevailing wisdom about how companies should compete in the new economy.

Conventional wisdom suggests that global companies should become "stateless." They should seek low-cost production wherever they can find it and build operations in many national markets. Yet SMH is committed to its Swiss home base. The bulk of its technology, people, and production remain anchored in the towns and villages around the Jura mountains on Switzerland's border with France—the traditional heart of Swiss watchmaking.

Companies also hear endless advice to become niche players by focusing on evermore-narrow market segments. But SMH is everywhere. Its brash and playful Swatch (whose basic models sell for $40) has become a pop-culture phenomenon. Last year, SMH sold an estimated 27 million Swatches. The company's flagship brand on the high end is Omega, whose models retail for anywhere between $700 and $20,000.

Watches in the hotly competitive middle segment sell for between $75 and $350. This is where SMH battles most directly with Seiko and Citizen, the two Japanese giants. SMH recently acquired Blancpain, a niche producer of luxury mechanical watches that retail for $200,000 and higher. SMH has even extended its presence in watches to new product categories such as telecommunications.

Finally, managers increasingly believe they should dismember their companies and retain only those core activities crucial to success. But SMH is a vertically integrated fortress. It assembles all the watches it sells, and it builds most of the components for the watches it assembles.

William Taylor, former HBR associate editor, conducted the interview at the Zurich headquarters of Hayek Engineering and at SMH headquarters in Biel.

HBR: The revitalization of SMH has generated tremendous attention across Europe. What are the most important lessons?

Nicolas Hayek: There are two main lessons. First, it is possible to build high-quality, high-value, mass-market consumer products in high-wage countries at low cost. Notice I said *build*, not just design and sell. A Swatch retails for 50 francs in Switzerland and $40 in the United States. The price has not changed in ten years. Yet we build all of our Swatches in Switzerland—where the most junior secretary earns more than the most senior engineer in Thailand or Malaysia.

In fact, it's not just possible to build mass-market products in countries like Switzerland. It's mandatory. This is a principle I am passionate about—and a principle business leaders in the United States and Europe don't take seriously enough. We are all global companies competing in global markets. But that does not mean we owe no allegiance to our own societies and cultures.

Not so long ago, I was in the United States for a meeting with the CEO of one of your big companies. We were discussing a joint venture to produce a new product we had developed. He saw what the product could do, he reviewed the design, and he got very excited: "Great, we'll make it in Singapore." His people had done no research or calculations at all. It was a reflex. I said, "No, we'll make it in Alabama."

But I'm sure you understand the reasoning. In a global economy, companies must source wherever they can get the best people at the lowest cost. Asia is bursting with smart, well-trained workers and engineers. How can you afford not to look there for manufacturing?

We must build where we live. When a country loses the know-how and expertise to manufacture things, it loses its capacity to create wealth—its financial independence. When it loses its financial independence, it starts to lose political sovereignty.

We have to change that reflex: the instinctive reaction that if a company has a mass-market consumer product, the only place to build it is Asia or Mexico. CEOs must say to their people: "We will build this product in our country at a lower cost and with higher quality than anywhere else in the world." Then they have to figure out how to do it.

We do this all the time. We agree on the performance specifications of a new product—a watch, a pager, a telephone. Then we assemble a project team. We present the team with some target economics: this is how much the product can sell for, not one penny more; this is the margin we need to support advertising, promotion, and so on. Thus these are the costs we can afford. Now go design a product and a production system that allows us to build it at those costs—in Switzerland.

That means focusing on labor. If we can design a manufacturing process in which direct labor accounts for less than 10% of total costs, there is nothing to stop us from building a product in Switzerland, the most expensive country in the world. Nothing.

But you know consumer markets are fiercely competitive. How can you absorb even a modest cost disadvantage?

This is not commodity competition. Let's say you have three similar watches. One says "Made in Japan" and sells for $100. Another says "Made in Switzerland" and sells for $110. A third says "Made in Hong Kong" and sells for $90. Which watch will consumers prefer? In Europe, between 75% and 95% of all consumers will prefer the Swiss watch—in spite of the 10% premium. In the United States, depending on which region you are talking about, between 51% and 75% of all consumers will prefer the Swiss watch. Only in Japan itself will a majority of consumers prefer the Japanese watch to the Swiss watch.

What does that mean? If you have a manufacturing process in which direct labor is less than 10% of total costs, you have eliminated those costs from the competitive equation. When we created SMH, our direct-labor costs, on average, were more than 30% of total costs. Today they are well below 10%. If we paid our workers full salaries and the Japanese paid their workers nothing, we could still compete.

This same logic applies beyond watches. CEOs must understand this point. If you can design a system in which direct-labor costs are less than 10% of total costs, it is cheaper to build mass-market consumer products in the United States than in Taiwan or Mexico.

What's the second lesson of SMH's comeback?

It's related to the first. You can build mass-market products in countries like Switzerland or the United States only if you embrace the fantasy and imagination of your childhood and youth. Everywhere children believe in dreams. And they ask the same question: Why? Why does something work a certain way? Why do we behave in certain ways? We ask ourselves those questions every day.

People may laugh—the CEO of a huge Swiss company talking about fantasy. But that's the real secret behind what we have done. It's an unusual attitude for Switzerland—and for Europe. Too many of Europe's large institutions—companies, governments, unions—are as rigid as prisons. They are all steel and cement and rules. We kill too many good ideas by rejecting them without thinking about them, by laughing at them.

Ten years ago, the people on the original Swatch team asked a crazy question: Why can't we design a striking, low-cost, high-quality watch and build it in Switzerland? The bankers were skeptical. A few suppliers refused to sell us parts. They said we would ruin the industry with this crazy product. But the team overcame the resistance and got the job done.

The Swatch is based on radical innovations in design, automation, and assembly, as well as in marketing and communications. One of our plants in Grenchen makes up to 35,000 Swatches and millions of components a day. From midnight until 8 A.M., it runs practically without human intervention. Swatch is a triumph of engineering. But it is really a triumph of imagination. If you combine powerful technology with fantasy, you create something very distinct.

Let's go back to the beginning of SMH—the dark days—and work forward to the two main lessons. It's the early 1980s. Swiss watchmaking is on the brink. The banks call you in. What do you find?

A chaotic jungle. An absolute mess. Most people who analyze the destruction of the Swiss watch industry in the 1970s emphasize price and technology. They point to the arrival of hundreds of millions of cheap quartz watches from Japan and Hong Kong and our decision to ignore quartz, a technology we invented. But we had huge problems beyond technology. There were problems of strategy, structure, management.

The two companies that became SMH were the flagships of the Swiss industry. One was SSIH, a company that had Swiss-French origins. Omega was the crown jewel of SSIH. Up until the early 1970s, Omega was one of Switzerland's most prestigious brands—more prestigious than even Rolex. But Omega was so successful for so long that it ruined SSIH. The company got arrogant. It also got greedy. It wanted to grow too fast, and it diluted the Omega name by selling too many watches at absurdly low prices.

SSIH had no discipline and no strategy. It had its own distribution in countries like Germany and France. It worked through agents in other countries. It even let some agents contract with outside manufacturers to build their own Omega models! It made no sense.

Then there was ASUAG, a company with Swiss-German origins. ASUAG was a manufacturing company. ASUAG owned a few brands, including Rado and Longines. But its heart and soul was an operation called Ebauches S.A., which supplied components to the whole Swiss watch industry. Ebauches was very capable. Over the years, however, as various small Swiss brands faltered, they went to ASUAG looking for a rescue package. ASUAG was the rich uncle: "Please, we are going broke. Why don't you buy us? It won't cost much. You can't let us disappear."

ASUAG felt a certain responsibility. Also, it didn't want to lose its customer base. By the late 1970s, then, ASUAG owned much more than Rado, Longines, and Ebauches. It owned all kinds of brands. It also owned lots of little companies that made components and had run into trouble. Many of these companies had been family-owned for generations. ASUAG rescued them but left most of the family managers in place.

By 1982, ASUAG owned more than 100 separate companies—some big, some small, some modern, some backward. Most of these companies did their own marketing, their own R&D, their own assembly. It was crazy.

The banks agree to merge SSIH and ASUAG to create SMH. It's time to sort through the jungle. Where do you begin?

We began with the products themselves. We had to understand our strategic positioning, where we stood in world markets. We made a big study, written in German, that became known as the Hayek Report. As you can imagine, the report got lots of attention in Switzerland. It was very controversial.

In the report, we drew a diagram to describe our competitive environment. It looked like a three-layer wedding cake. Back then, the world market for watches was about 500 million units per year. The low-end segment, the bottom layer of the cake, had watches with prices up to $75 or so. That layer represented 450 million units out of 500 million. The middle layer, with watches up to $400 or so, represented about 42 million units. That left 8 million watches for the top layer, with prices from $400 into the millions of dollars.

The Swiss share of the bottom layer, 450 million watches, was zero. We had nothing left. Our share of the middle layer was about 3%. Our share of the top layer was 97%.

We were cornered. The Swiss spent much of the 1970s reacting to quartz by retreating: "Why should we compete with Japan and Hong Kong? They make junk, then they give it away. We have no margin there." Of course, as we retreated, the Japanese moved up to the next layer of the cake. Then the retreat would start again.

I decided we could retreat no longer. We had to have a broad market presence. We needed at least one profitable, growing, global brand in every segment—including the low end. That explained why we had to control and sell Swatch ourselves rather than license it or sell it through agents, as some people had proposed. It also meant we had to reinvigorate Tissot, the only global brand we had in the middle segment.

The banks studied our report and got nervous, especially about Swatch. Some of them worried that Swatch would cannibalize Tissot. Others were more emphatic: "This is not what consumers think of when they think of Switzerland. What the hell are you going to do with this piece of plastic against Japan and Hong Kong?" But we were adamant: if we did not have mass production, if we did not have a strong position in the low end, we could not control quality and costs in the other segments.

The banks thought Swatch would fail?

Not fail, necessarily. Many thought it might survive—but barely. After all, we were going to battle at the low end. Who could make real money there against people from Japan or Hong Kong? That's when they proposed that I buy 51% of the company.

What did you see that others didn't?

I understood that we were not just selling a consumer product, or even a branded product. We were selling an emotional product. You wear a watch on your wrist, right against your skin. You have it there for 12 hours a day, maybe 24 hours a day. It can be an important part of your self-image. It doesn't have to be a commodity. It shouldn't be a commodity. I knew that if we could add genuine emotion to the product, and attack the low end with a strong message, we could succeed.

How do you "emotionalize" a watch? Do you mean to say that Swatch turned something that was mundane and functional into a fashion statement?

That's how most people describe what we did. But it's not quite right. Fashion is important. The people at our Swatch Lab in Milan and our many other designers do beautiful work. The artists who make our Swatch special collections design wonderful watches (see "Franco Bosisio: In the Eye of the Swatch Storm"). But take a trip to Hong Kong and look at the styles, the designs, the colors. They make pretty watches over there too.

We are not just offering people a style. We are offering them a message. This is an absolutely critical point. Fashion is about image. Emotional products are about message—a strong, exciting, distinct, authentic message that tells people who you are and why you do what you do. There are many elements that make up the Swatch message. High quality. Low cost. Provocative. Joy of life. But the most important element of the Swatch message is the hardest for others to copy. Ultimately, we are not just offering watches. We are offering our personal culture.

Franco Bosisio: In the Eye of the Swatch Storm

From his office in Milan, Franco Bosisio sits in the eye of the Swatch storm. He established and now runs the main Swatch Design Lab, an ever-changing collection of artists, architects, and industrial designers who turn out many of the 140 new Swatch models every year. Bosisio also runs SMH Italy, where the Swatch phenomenon has reached its most fevered pitch.

"It's almost embarrassing. I go out to dinner, and all I see are Swatches. Swatch now accounts for 40% of the watches sold in Italy. But if you combine the watches we sell here, plus the watches Italians buy when they travel abroad, plus what they buy through auctions and 'unofficial' channels, our market share is even higher.

"It's incredible! And I can't really explain it. We didn't launch Swatch here until 1986. Italy was the last major European country to get it. That's strange, in a way, because Italians are crazy for fashion in general and watches in particular. So people were dreaming about it: When are we going to get Swatch? When we finally made the launch, the market exploded. And it has kept growing. Our average customer in Italy owns six Swatches.

"This was no brilliant strategy on our part. For years, we just weren't happy with the distribution situation in Italy. So we decided to wait until we could do it right. Then came a series of high-profile events. For example, Sotheby's chose Milan as the site of one of its first Swatch auctions. None of us knew what to expect. But the frenzy and the prices and the visibility were incredible. People were pushing their way in, their pockets bulging with bank notes. It was really emotional. My public relations person was in tears because it was so intense. The people from Sotheby's were astonished.

"So luck has played a big role in our success here. But of course there is more. I have always thought the appeal of Swatch rests on four pillars: design, communication in the widest sense, quality, and price. Few people appreciate how and why price has been so important. Everywhere in the world, Swatch is sold at an affordable price. But it's also a simple price, a clean price. In the United States, $40. In Switzerland, SFr50. In Germany, DM60. In Japan, ¥7,000.

"It has also been, in the first ten years, an unchanging price. Despite our incredible success in the market, despite the huge unmet demand in countries like Italy, we have never raised the price of Swatch to our dealers. And we forbid our dealers from marking up the price to their customers. Can you name another fashion product whose price has stayed exactly the same for ten years?

"Price becomes a mirror for the other attributes we try to communicate. It helps set us apart from the rest of the world. A Swatch is an easy decision to make, an easy decision to live with. It's provocative. But it doesn't make you think too much.

"That's also the sensibility behind our designs. There are about 20 people in our Swatch Lab. They come from many backgrounds and from all over the world. We have had designers from Italy, Japan, Germany, France, America, Australia, many other places. A few are trained in architecture. Many are trained in industrial design. A few are right out of university.

"People work with us for many reasons. Let's face it, most people who are serious about a career in design want to spend some time in northern Italy. And they love the product, the idea that millions of people wear it. They also value the chance to interact with Alessandro Mendini, our art director. Mendini is a major figure in European styling. His work occupies a space somewhere between fine art and industrial design. He has long been known for his dramatic furniture, which he makes in very small numbers. Swatch is really his first experience with mass production.

"By the way, Mendini has managed to combine both interests. He designed Lots of Dots, a special Swatch only available to members of our Swatch collector's club. Then he designed a ceramic armchair that is based on Lots of Dots.

"We rotate people through the lab. Everyone must agree to stay for at least six months. Very few people stay for more than two years. They get tired. We get tired. This is challenging work. Lots of brilliant artists fail when they try to design for us. It's hard to squeeze wonderful ideas into the face and band of a Swatch. In fact, we have more than 3,000 proposed designs that we have never executed.

"We design two Swatch collections each year, 70 styles per collection— 140 different models. We always prepare four times the number of models we need. A group in Milan selects half of those proposed designs and presents them to the management committee in Switzerland. In Switzerland, we make the final choices and decide on the collection.

"Four years ago, when we established the Lab, I worried that we would run out of ideas. We asked a group of fashion experts—famous journalists, art dealers, and so on—to prepare reports for us on big themes. We worked with Oliviero Toscani, the photographer who does much of Benetton's advertising. We hired all kinds of consultants.

"We still do that—but we now do much, much more. These experts are so accomplished that they are asked to advise lots of other companies. So it was hard for us to be distinctive. We also came to realize how repetitive fashion can be. With clothing, you change the shape, the length, the fabric. But the basic design elements—the decorations and patterns— don't change nearly as much. For us, of course, it's the exact opposite. The size and shape of the watch stay the same. What varies is what's on it.

"We have no set routines to come up with ideas. We travel constantly, all over the world. We go to the big fashion shows—Première Vision in Paris, for example. We go to the opera, to art exhibitions. You can't imagine how many books and magazines we read, how many painters we study. We steep ourselves in the culture of life. And then things happen.

"Let me give you an example. Two of our watches in the latest Fall/Winter collection were called Fairy Tales. They grow out of some books we came across, books of images that inspired some of the great artists of the nineteenth century. On one of the watches there is a teapot, there is a man kissing the hand of a woman, and there is the face of a young woman. The holes in the bracelet look like tears. That watch tells a story. It's like a chance encounter with people from an earlier time.

"We don't get carried away. This is only semi-serious. We want arresting images, but we also wink at the consumer. We don't want people to think too much. We are looking for an immediate emotional reaction—spontaneity. All we can tell is that there is a man romancing a woman and that the woman is crying. Everything else is left to your imagination."

You mean people can look at a watch from Hong Kong and then look at a Swatch and sense a different culture?

It's not just the mechanics of the product. It's also the environment around the product. One thing we forget when we analyze global competition is that most products are sold to people who share our culture. Europeans and Americans are the biggest groups in the world buying products from Asia. So if you can surround your product with your own culture—without ever denigrating other cultures or being racist in any way—it can be a powerful advantage.

We are offering our products to a sympathetic audience. The people who buy Swatches are proud of us. They root for us. They want us to win. Europeans and Americans are damn happy if you can show that their societies are not decadent—that every Japanese or Taiwanese worker is not ten times more productive or more intelligent than they are.

What is a concrete example of embedding culture in the message around Swatch?

How did we launch Swatch in Germany? Did we saturate the airwaves with paid advertisements? No. Anyone can do that. We built a giant Swatch. It was 500 feet high, weighed 13 tons, and actually worked. We suspended that giant Swatch outside the tallest skyscraper in Frankfurt, the headquarters of Commerzbank. It was really something to see!

I remember asking the chairman of the bank for permission. He thought we were crazy. We were crazy, but we had already gotten authorization from the city engineers and the local government. And we persuaded him that this giant Swatch would show his customers that his bank had heart and emotion. So there it hung. And all it said was: Swatch. Swiss. DM60.

That outrageous display communicated the essence of the Swatch message. It was high quality—Swiss. It was low cost—what could be more affordable than DM60? It was a big provocation to hang a watch from a huge, grim skyscraper. And it was funny, fanciful, a joke—joy of life. Believe me, when we took it down, everyone we had wanted to reach had received our message.

We also hung a giant Swatch in Tokyo, in the Ginza. This message can work in Japan as well. By value, Swiss companies account for more than 50% of all the watches sold in Japan. SMH accounts for 75% of that 50%. Do you think we broadcast these figures? Or that we act arrogantly in Japan? Of course not. The Japanese are sympathetic to us. We're nice people from a small country. We have nice mountains and clear water. They like us and our products, and we like them.

So you're really talking about advertising and promotion?

I am talking about everything we do. Everything we do and the way we do everything sends a message. Let me give another example—our "Swatch the World" celebration in Zermatt last fall. This was a huge occasion for us. It celebrated the production of our 100 millionth Swatch. It was a very public event. Political leaders came to Zermatt. There was lots of hoopla, lots of excitement.

We could have spent millions of francs on champagne, caviar, and dancing. Instead, we created an artistic event, a tasteful event, an honest event. It was set in a village at the foot of the Matterhorn, the symbol of Swiss authenticity. But it was lively and offbeat. We had street musicians, artists, reggae, clowns. We also had Jean-Michel Jarre, the brilliant young composer from France. He wrote original music for us and beamed a multimedia light show against the Alps.

More than 40,000 people came to Zermatt from all over Europe. The newspapers and magazines and television stations carried big reports. It was a critically important statement of the Swatch message. And what were we communicating? Our culture. Our sense of ourselves and what we stand for.

Do you use the same logic with all of your other brands too?

Each brand is different, so each message is different. But each brand has a message. My job is to sit in the bunker with a machine gun defending the distinct messages of all my brands. I am the custodian of our messages. I review every new communications campaign for every single brand.

Let's talk about Omega. We built Swatch from the ground up. We had to rescue Omega from near oblivion. Omega is a crucial brand for SMH. We make as much money with Omega as we do with Swatch. Yet no one ever discusses it.

So why was Omega near oblivion?

Omega is one of the Swiss watch industry's great brands. Its history goes back to 1848. You should visit the watchmaking museums and look at the pieces Omega made 50 or 100 years ago. They are wonderful. Few brands had or have Omega's potential power.

The problems started in the early 1970s. There were bad business practices. The people there became arrogant. They treated their agents and dealers badly. If an agent from, say, New Jersey needed 200 units of a particular model, Omega would say: "You're crazy! Don't bother us with such nonsense. We'll give you 50."

Second, and much worse, Omega became greedy. Rolex sells 600,000 watches per year. That's about as many as you can sell before a luxury brand begins to lose its prestige. That's about how many Omega was selling in the late 1970s. But Omega wanted to grow more rapidly. So they took the easy route. They figured, "If we can sell 600,000, why not a million? Or 2 million? Or 3 million?"

Which meant, of course, they had to lower the price radically. A jeweler would say, "Omega is wonderful, but it is too expensive for my clients. How about giving me an Omega that is cheaper?" Now, if you are crazy, or I guess if you are greedy, you agree.

That was the kiss of death. Omega was everywhere: high price, medium price, precious metals, cheap gold plating. There were 2,000 different models! No one knew what Omega stood for. By the end of 1980, the company was again in a deep crisis, its deepest ever.

How did you develop the recovery plan?

It was very painful. Early on, before the SMH merger, we spent days behind closed doors with the company's managers and bankers. Some of the people suggested that we sell Omega to the Japanese, who had offered to buy it. It was still a powerful brand despite the problems, and it would fetch a lot of money. Of course, that would have been tragic for Switzerland.

Some managers from Omega were pleading that we take the brand down-market to face Citizen and Seiko in the middle segment. That was absurd! I fought them tooth and nail on that. At one point, the people at Omega would not even let us on their premises. It was bitter.

We made the investment in SMH in 1985, and we were free to act. I pushed out practically the entire management of Omega. I fired a lot of people. I got the reputation as a brutal guy. I am not a brutal guy. But the organization was so full of arrogance and stupidity that I didn't have much choice. Then we implemented our strategy to give Omega back its message. Not an image, a message. We had to be clear, consistent, and credible.

What is Omega's message?

Omega is an elite watch for people who achieve—in sports, the arts, business, the professions—and help shape the world. It is a watch for people who are somebody because they made themselves somebody, not because their grandfather left them a trust fund or because they made money from insider trading. The astronauts who landed on the moon achieved something. They were smart, healthy, courageous. They wore Omega. So did the Soviet cosmonauts. That message had been destroyed.

You also reorganized manufacturing, assembly, and distribution. Did these changes play a big role in Omega's turnaround?

We did all those things. We paid closer attention to logistics and quality and inventories. We redesigned the manufacturing process. Omega gave up most of its manufacturing when we created SMH and centralized production. There was a lot of resistance. People said we would destroy the brand.

We do treat Omega specially, however. ETA makes movements for Omega as it does for all our brands. But we sell those movements to no one else inside or outside the company. Because those movements are specially made, ETA stamps the Omega name, not ETA, on the movement. This way, when customers or dealers open a watch to repair it or change the battery, they still see Omega. By the way, this is true only for quartz watches. Omega designs most of its own mechanical movements and buys them from Frédéric Piquet, as do most high-class Swiss watches. We own Piquet.

All these reforms were important. But they're not the main story. The main story is the message and the discipline the message imposes. We immediately reduced the number of models from 2,000 to 1,000. We have since cut back to 130. We don't allow anyone to make Omegas under license. We stopped building lavish or showy watches—people who achieve don't care about those things. We stopped making Omegas with gold plating. We make smart watches out of real metal: platinum, titanium, gold, special steel alloys.

Omega started making sense again. This is what an Omega looks like. This is what Omega stands for. We gave Omega back its message.

Swatch and Omega have completely different messages. Why does it matter strategically that SMH owns both brands? How does the success of Swatch help Omega?

It's not just Omega. Swatch has helped the entire Swiss industry. It has restored our credibility with the public. It has restored our credibility with the trade. The perception of Swiss watches today is so different from ten years ago. Jewelers and department stores beg us to carry Swatch. If customers come for Swatch, and the shop doesn't have it, they won't accept anything else. They leave disappointed. This has a big psychological spillover to all of our brands—including Omega.

I really believe the phenomenal success of these $40 watches helps the climate for selling $500 watches—or $5,000 watches, for that matter. We have re-established our technical superiority over the Japanese watchmakers. If we can build beautiful, high-quality watches that sell for only $40, imagine what must be the quality and accuracy of watches that sell for $2,000!

Let's move from products to production systems. This is the age of the lean corporation. Why is SMH so vertically integrated?

We are vertically integrated because it is the only way to maintain our strategic independence and freedom to maneuver in the market. We can't be broad in the marketplace without being deep in our production systems. And we can't support deep production systems without a broad presence in the marketplace.

Here is the strategic reality we face. There are three centers of watchmaking in the world: Switzerland, Japan, and Hong Kong. But only the Swiss and the Japanese build movements, the complex micromechanical and electronic components that are the "guts" of a watch. To build electronic movements, you have to master semiconductor technology, quartz, batteries, miniaturization. For mechanical movements, you have to master intricate micromechanics and the old art of watchmaking in the luxury segment. You either commit fully to the business or get out.

We have some advantages here in Switzerland. We have hundreds of years of experience in the technologies and techniques of watchmaking. Families have spent generations in our factories. They have a feel for this business, a special touch. They are masters at working with parts so small you need a microscope to see them. Not to mention the tooling to build these parts. There are plenty of toolmakers around the world, but not for parts with these dimensions.

When we designed the Swatch factory, we built special machines for injection molding, automated assembly—virtually everything. There were only a handful of people in the world with the know-how to build those machines. They all lived in this part of Switzerland.

If you had all this expertise, why was the company so troubled?

We spent much of the 1980s reinventing our production systems to leverage our expertise. I've already described the mess we found at ASUAG and SSIH. So we made massive reorganizations. Brands no longer build their watches. Our production divisions have full responsibility for manufacturing and assembly. We standardized as many parts as we could among the brands: quartz resonators, stepping motors, and so on. We standardized much of the tooling for these parts. We made huge investments in high-tech operations like semiconductors and batteries. We made huge commitments to automating our assembly lines.

Our cost structure today is completely different from when we created SMH. It is not merely because we are more productive. It is because we have totally changed the logic of our business system. We have radically decentralized marketing and thoroughly centralized manufacturing. There is no comparison between what exists at SMH today and what existed at ASUAG and SSIH (see "The Many Faces of SMH").

The Many Faces of SMH

Nicolas Hayek runs a competitive juggernaut—one of Europe's most powerful and well-positioned companies. It is also a company of staggering complexity. SMH employs 15,000 people. It sold 100 million finished watches and movements in 1992. With an average of 150 to 200 parts for each watch or movement, this sales volume translates into 15 billion to 20 billion parts. SMH makes most of them, assembles all of them, and ships them to places like Chicago or Buenos Aires or Hanoi.

There are other companies in other industries that cover every market segment and make and assemble most of their own parts: General Motors in automobiles, IBM in computers, Philips in consumer electronics. What these companies have in common is that they are struggling to reinvent themselves in a world in which the power of integration gives way to the nimble flexibility of innovation. Increasingly, scale and success seem to be at odds.

But this is not true at SMH. Why? Because the company has resolved a set of tensions that would cripple most organizations.

Think of the opportunities for misunderstanding and conflict within

SMH. There are the language and cultural tensions between workers and managers with roots in the Swiss-German ASUAG and those with roots in the Swiss-French SSIH. There are tensions between SMH's fierce style of competition in Hong Kong, where it wages hand-to-hand combat over millions of $5 watch movements, and the genteel world of Blancpain, which sells a few thousand watches each year for $200,000 and higher. There are tensions between ETA's disciplined production engineers in Switzerland and the free-spirited designers in the Swatch Lab in Milan.

SMH resolves these tensions by creating clear-but-flexible organizational boundaries and encouraging what Hayek calls "conflict of targets." Each unit of SMH operates with clear responsibilities and fixed rules. But no unit can do its job without interacting and negotiating with others. In this respect, SMH looks more like a self-contained industry than a single company. The goal is to surface and resolve differences quickly and without bureaucratic rancor.

For example, Hayek has radically decentralized marketing. SMH has nine major global brands: Blancpain, Omega, Longines, Rado, Tissot, Certina, Hamilton, Swatch, and Flik Flak. Each brand has its own organization, its own management even its own building. Brands have total authority over design, marketing, and communications. But they play a sharply limited role in manufacturing or assembly. For this, they must negotiate with SMH's technology and manufacturing arms.

One critical manufacturing organization is ETA, which builds all of SMH's quartz movements. ETA, which is based in Grenchen, 20 minutes south of SMH headquarters in Biel, is responsible for all electronic-movement research, manufacturing, and assembly. ETA is a production empire. It runs giant factories that build movements for most of the Swiss watchmakers. It operates highly automated assembly plants throughout the country as well as in Thailand. Of course, none of this matters unless ETA can satisfy its main customers—the SMH brands. So there are ongoing negotiations over style versus quality, design versus manufacturability, speed versus cost.

These strict boundaries and tough negotiations allow SMH to minimize overhead. The company has remarkably few bureaucratic monitors or headquarters referees. Instead, near the top of SMH sits a 16-member Enlarged Group Management Board. The executives on this board work around the world and have direct responsibility for the company's many business units. They include Anton Bally, the powerful head of ETA; Raymond Zeitoun, who works in New York and runs all SMH operations in the United States; Florence Ollivier-Lamarque, who runs SMH France from Paris; Hans-Jürg Schär, responsible for Swatch; and Hans Kurth, who has worldwide responsibility for Omega. (A seven-member Restricted

Group Management Board, whose members are drawn from the larger group, is the highest SMH management council.)

When these 16 executives convene, as they do once a month (the restricted board meets twice a month), they share information from key national markets, surface and resolve disagreements among competing business units and geographies, and then return to their organizations with the company's newest game plan. In other words, there are minimal delays between decision and action.

Finally, there is Nicolas Hayek. As CEO, Hayek injects himself directly and visibly throughout SMH—especially on issues of strategy and new products. He receives almost daily briefings and progress reports on new products—watches, pagers, telephones, and others—and constantly works with (and pressures) managers and engineers to cut through red tape. He even has final signature approval over every design in every Swatch collection.

The end result is a company that is solid without being stodgy: a tough, fast-moving competitor whose remarkable turnaround demonstrates the power of SMH's unconventional management principles.

SMH has invested more than SFr1 billion in technology and factories. Why such enormous investments?

We have been making up lost ground. Throughout the 1970s, as companies in Hong Kong assembled hundreds of millions of cheap watches, Japan made investments to supply them with movements. They built big factories and slashed prices. A movement used to cost, on average, between $8 and $20. Today it costs between $2 and $5. The Swiss couldn't play that game. We sold movements only to very special people in Hong Kong—companies building watches for brands like Timex, who wanted to market to customers in the United States that their watches had Swiss movements. We were uncompetitive on cost. We just didn't have the volume.

So we decided to get much more aggressive. We had to confront the Japanese in Hong Kong. We seized every opportunity we could. One day, for example, I was seated with an entrepreneur from Hong Kong who sold finished watches. He bought all his movements from one of the giant Japanese companies. He was begging me for an Omega agency in Asia. I said, "How can you expect me to give you an Omega agency when you buy 20 million movements a year from a competitor of mine?" He told me, "I will buy from you if you give me the same price."

I agreed to the order on the spot because the price was profitable for us. ETA worked night and day to produce those movements. But it was an important opportunity for us. We could not allow the Japanese to control movements in Hong Kong. They would have developed enormous leverage over us.

What do you mean?

Imagine if we got so uncompetitive in movements that we had to get out of the business altogether. And imagine if, like so many companies in other industries, we then focused on design and marketing and final assembly. And then imagine if we designed a great new model for Tissot, a watch that was going to be a big hit in the United States and hurt our rivals.

Now, what if our Japanese supplier recognized the appeal of that new model? You can imagine what might happen: "We are very sorry, but we are having capacity problems at our plant. We know we used to charge you $3.50 for that movement. It is now $12." Well, that destroys the economics of the watch. So the next thing the supplier says, "Of course, we can make special arrangements in our plant. But we prefer to deal with partners rather than customers. We are prepared to become 50% owners of the brand and supply movements at more favorable prices."

That sounds conspiratorial. We heard that argument about Japanese leverage in computers because they dominate flat-panel displays and power supplies. The Japanese compete fiercely among themselves, why should they conspire against European companies?

Let me correct myself. It is wrong to talk about "the Japanese" in our business. In movements, there are only two Japanese companies that really matter—Citizen and Seiko. The moment you get five or ten Japanese companies in the same industry, they do fight like hell. But there are plenty of industries where that is not the case, and ours is one.

By the way, I am not suggesting that movement suppliers would single out Europeans or Americans. We sell lots of movements to watch companies in Japan. Why? Because Citizen and Seiko sometimes can't or won't supply them. We sold movements to Casio when it started making watches. Why? Because Citizen and Seiko couldn't or wouldn't supply them. That's the reality.

You also invested hundreds of millions of Swiss francs to make semiconductor chips. But the world is flooded with chips. Isn't that a case of unnecessary vertical integration?

No, it isn't. The world is flooded with five-volt and three-volt circuits—the chips used in computers, televisions, and VCRs. But watches use 1.5 volt chips. There are three companies in the world that dominate production of 1.5 volt chips. One is SMH. The other two are in Japan. We must make our own chips to maintain our independence.

We considered all the options before we made this investment. Some of our people wanted to close down the operation and find other companies to build chips for us. We visited companies from Norway to the United States. We showed them our designs and said we would cooperate with them. The best quotes were three times our costs. So that didn't work.

Some of our people wanted us to diversify into computer chips as well, but that was crazy. How would competing in memories or microprocessors help us in watches? And how were we going to go up against companies such as Intel or Motorola or Toshiba?

So we decided to make our own chips and to make them in huge volumes. That meant mass markets—Swatch. It meant pushing harder on movement sales—Hong Kong. It also meant creating new markets for 1.5 volt chips. Today you can find our chips in hearing aids, pacemakers, cellular telephones, even the monitor switch for antilock brakes. In fact, more than 50% of our revenues from 1.5 volt chips come from markets outside the watch industry. Those revenues and profits help us keep investing. And they also keep volumes up and costs down.

Can't you take this too far, though? Why risk diversifying into products like pagers and telephones and building them yourself?

What risk are you talking about? Our pagers grow directly out of our experience with watches. There is tremendous proprietary technology in the Swatch Pager. It uses high-precision injection molding that requires specialized tooling—just like the Swatch. It uses very advanced stepping motors, a technology we mastered in our watches. Who is better suited for this market than we are?

Look at the wristwatch pagers out there today. They have digital displays. They are ugly. They are heavy. It's no accident we will market the world's first successful analog wristwatch pager. We have the technology. We know how to design manufacturing systems that make us competitive. We also know how to design attractive products—and how to market them.

How many worldwide brands can you name for telecom products sold directly to the end customer? Alcatel or Siemens understand how to sell to governments and big companies. We understand how to sell to people. This is going to be a big business for us.

I understand the strategic case for integration. But how do you guard against the bureaucracy it always seems to create?

We don't need elaborate planning bureaucracies or corporate staffs. If we have a big decision, we assemble the best group we can and let them go at it. Then we move on to other things.

Organizational structure is the most inhuman thing ever invented. It goes against our nature as people. So we have clear boundaries and targets. Our brands work independently of one another. The people at Omega and Rado and Tissot have their own buildings. They have their own managements. They are responsible for their own design, marketing, communications, and distribution. They are emotionally connected to their brands, not just to SMH as an entity. I want people at Rado to love Rado. And I want people at Longines to love their brand.

We are big believers in decentralization. This company has 211 profit centers. We set tough, demanding budgets for them. I personally participate in detailed budget reviews for our major profit centers. Then we track performance closely. We get monthly sales figures for all profit centers on the sixth day of the following month. We get P&L statements about 10 or 15 days later. The moment anything looks strange, we react very quickly, very decisively, very directly.

And we are big believers in project teams. Find your best people, let them tackle a problem, disband them, and move on to the next problem. I suppose this comes from my consulting background. Every company needs certain functions. You have to watch costs every day. You have to watch quality every day. But you don't decide every day whether to launch a Swatch Pager or whether to build a new chip plant.

Of course, this approach works only if the whole management team is totally focused on developing products and improving operations, not fighting with each other. Last September, Goldman Sachs in London issued a report on our company. It said our management is "obsessed with productivity." We're not obsessed. But we do keep our eye on what matters. We don't spend a penny that we don't have to spend.

Is SMH corporation a realistic model for other companies to emulate?

Everything we've done can be done by lots of companies in Switzerland, or France, or Germany, or America. All it takes is the will to do it. Which is, I admit, no small matter. As a consultant, I had been preaching, shouting, writing for years about how European companies could compete with the Far East. At SMH, I got the chance to practice what I preached. But there were some special factors at work here.

For one thing, the Swiss watch industry was completely devastated. People had given up; they were ready to sell our most valuable brands to foreign competitors and sell our factories for their real estate value. If ASUAG or SSIH had been making one franc of profit, they would have thrown me out the window: "You're crazy, what do you know about watches anyway?"

Second, I put my own money on the line, along with money from our investors. The fact that our group controls a majority of the equity means we could make decisions that other people were scared to make.

I inflicted pain, made controversy, created worry. Nobody believed the targets we set back in the mid-1980s. People thought we were crazy to invest SFr1.1 billion in one of the highest cost regions in the world. But it was our investment on the line. There was never any doubt who was running the show—or, for that matter, who would get blamed if we failed.

PART

II

Designing
New Behaviors

1
Research That Reinvents the Corporation

John Seely Brown

The most important invention that will come out of the corporate research lab in the future will be the corporation itself. As companies try to keep pace with rapid changes in technology and cope with increasingly unstable business environments, the research department has to do more than simply innovate new products. It must design the new technological and organizational "architectures" that make possible a continuously innovating company. Put another way, corporate research must reinvent innovation.

At the Xerox Palo Alto Research Center (PARC), we've learned this lesson, at times, the hard way. Xerox created PARC in 1970 to pursue advanced research in computer science, electronics, and materials science. Over the next decade, PARC researchers were responsible for some of the basic innovations of the personal-computer revolution—only to see other companies commercialize these innovations more quickly than Xerox. (See "PARC: Seedbed of the Computer Revolution.") In the process, Xerox gained a reputation for "fumbling the future," and PARC for doing brilliant research but in isolation from the company's business.

That view is one-sided because it ignores the way that PARC innovations *have* paid off over the past 20 years. Still, it raises fundamental questions that many companies besides Xerox have been struggling with in recent years: What is the role of corporate research in a business environment characterized by tougher competition and non-stop technological change? And how can large companies better assimilate the latest innovations and quickly incorporate them in new products?

PARC: Seedbed of the Computer Revolution

Former Xerox CEO C. Peter McColough created the Palo Alto Research Center (PARC) in 1970 to perform basic research in computing and electronics and to study what McColough called "the architecture of information"—how complex organizations use information. PARC hired some of the best computer scientists in the world and gave them virtually unlimited funding to pursue their ideas.

The scientific payoff from PARC was immediate. Throughout the 1970s, PARC researchers produced a series of fundamental innovations in computer technology that would prove to be the building blocks of the personal-computer revolution: "bit map" display computer screens that make easy-to-use graphic interfaces possible, local area networks for distributed computing, overlapping screen windows, point-and-click editing using a "mouse," and Smalltalk, the first object-oriented programming language.

Xerox never became a dominant player in the personal-computer industry. But PARC's research has nevertheless directly fed the company's strategic business. PARC developed the first prototype of laser printing in 1973. By 1990, laser printing was a several-billion-dollar business at Xerox. And PARC's innovations in local area networks and its distinctive computer interface designs have been successfully incorporated in Xerox copiers and printers, an innovation that was crucial to the company's successfully meeting the challenge from Japanese competition in the 1980s.

Where PARC scientists of the 1970s had a technical vision, today the center is increasingly focusing on the interrelationships between technology and work. In 1990, anthropologists, sociologists, linguists, and psychologists complement PARC's traditional research staff of computer scientists, physicists, and engineers. And much of the center's computer-science research emphasizes how information technology can be used to support effective group collaboration—a field known as computer-supported co-operative work.   —Robert Howard

One popular answer to these questions is to shift the focus of the research department away from radical breakthroughs toward incremental innovation, away from basic research toward applied research. At PARC, we have chosen a different approach, one that cuts across both of these categories and combines the most useful features of each. We call it pioneering research.

Like the best applied research, pioneering research is closely connected to the company's most pressing business problems. But like the best basic research, it seeks to redefine these problems fundamentally in order to come up with fresh—and sometimes radical—solutions. Our emphasis on pioneering research has led us to redefine what we mean by technology, by innovation, and indeed by research itself. Here are some of the new principles that we have identified.

1. Research on new work practices is as important as research on new products. Corporate research is traditionally viewed as the source of new technologies and products. At PARC, we believe it is equally important for research to invent new prototypes of organizational practice. This means going beyond the typical view of technology as an artifact—hardware and software—to explore its potential for creating new and more effective ways of working, what we call studying "technology in use." Such activities are essential for companies to exploit successfully the next great breakthrough in information technology: "ubiquitous computing," or the incorporation of information technology in a broad range of everyday objects.

2. Innovation is everywhere; the problem is learning from it. When corporate research begins to focus on a company's practice as well as its products, another principle quickly becomes clear: innovation isn't the privileged activity of the research department. It goes on at all levels of a company—wherever employees confront problems, deal with unforeseen contingencies, or work their way around breakdowns in normal procedures. The problem is, few companies know how to learn from this local innovation and how to use it to improve their overall effectiveness. At PARC, we are studying this process of local innovation with employees on the front lines of Xerox's business and developing technologies to harvest its lessons for the company as a whole. By doing so, we hope to turn company size, so often seen as an obstacle to innovation, into an advantage—a rich seedbed of fresh insights about technology and new work practices.

3. Research can't just produce innovation; it must "coproduce" it. Before a company can learn from the innovation in its midst, it must rethink the process by which innovation is transmitted throughout the organization. Research must "coproduce" new technologies and work practices by developing with partners throughout the organization a shared understanding of why these innovations are important. On the one hand, that means challenging the outmoded background assumptions that so often distort the way people see new technologies, new market opportunities, and the entire business. On the other, it requires creating new ways to communicate the significance of radical innovations. Essentially, corporate research must prototype new mental models of the organization and its business.

4. The research department's ultimate innovation partner is the customer. Prototyping technology in use, harvesting local innovation, coproducing new mental models of the organization—all these activities that we are pursuing inside Xerox are directly applicable to our customers as well. In fact, our future competitive advantage will depend not just on selling information-technology products *to* customers. It will depend on coproducing these products *with* customers—customizing technology and work practices to meet their current and future needs. One role of corporate research in this activity is to invent methods and tools to help customers identify their "latent" needs and improve their own capacity for continuous innovation.

At PARC, we've only begun to explore the implications of these new principles. Our activities in each of these areas are little more than interesting experiments. Still, we have defined a promising and exciting new direction. Without giving up our strong focus on state-of-the-art information technologies, we are also studying the human and organizational barriers to innovation. And using the entire Xerox organization as our laboratory, we are experimenting with new techniques for helping people grasp the revolutionary potential of new technologies and work practices.

The result: important contributions to Xerox's core products but also a distinctive approach to innovation with implications far beyond our company. Our business happens to be technology, but any company—no matter what the business—must eventually grapple with the issues we've been addressing. The successful company of the future must understand how people really work and how technology can help them work more effectively. It must know how to create an environment for continual innovation on the part of *all* employees. It must rethink traditional business assumptions and tap needs that customers don't even know they have yet. It must use research to reinvent the corporation.

Technology Gets Out of the Way

At the foundation of our new approach to research is a particular vision of technology. As the cost of computing power continues to plummet, two things become possible. First, more and more electronic technology will be incorporated in everyday office devices. Second, increased computing power will allow users to tailor the technology to meet their specific needs.

Both these trends lead to a paradoxical result. When information technology is everywhere and can be customized to match more closely the work to be done, the technology itself will become invisible. The next great breakthrough of the information age will be the disappearance of discrete information-technology products. Technology is finally becoming powerful enough to get out of the way.

Consider the photocopier. Ever since Chester Carlson first invented xerography some 50 years ago, the technology of photocopiers has been more or less the same. In a process somewhat similar to photography, a light-lens projects an image of the page onto a photoreceptor. The image is then developed with a dry toner to produce the copy. But information technology is transforming the copier with implications as radical as those accompanying the invention of xerography itself.

Today our copiers are complex computing and communications devices. Inside Xerox's high-end machines are some 30 microprocessors linked together by local area networks. They continually monitor the operations of the machine and make adjustments to compensate for wear and tear, thus increasing reliability and ensuring consistent, high copy quality. Information systems inside our copiers also make the machines easier to use by constantly providing users with information linked to the specific task they are performing. (See "How Xerox Redesigned Its Copiers.") These innovations were crucial to Xerox's success in meeting Japanese competition and regaining market share during the past decade.

How Xerox Redesigned Its Copiers

In the early 1980s, Xerox's copier business faced a big problem. Service calls were increasing, and more and more customers were reporting that our newest copiers were "unreliable." The complaints couldn't have come at a worse time. We had been late to recognize market opportunities for low- and mid-range copiers, and Japanese competitors like Canon were cutting into our market share. Now Xerox's reputation for quality was at stake.

After interviewing some customers, we discovered that unreliability was not the real problem. Our copiers weren't breaking down more frequently than before; in fact, many of the service calls were unnecessary.

But customers were finding the copiers increasingly difficult to use. Because they couldn't get their work done, they perceived the machines as unreliable.

The source of the problem was our copier design. Traditionally, Xerox technology designers—like most engineers—have strived to make machines "idiot proof." The idea was to foresee in advance all the possible things that could go wrong, then either design them out of the system or provide detailed instructions of what to do should they occur.

But as we kept adding new functions, we had to add more and more information, usually stored on flip cards attached to the machine. The copiers became so complex that it was harder for the new user to figure out how to do any particular task. To learn a new operation meant a time-consuming search through the flip cards. And whenever something went wrong—a paper jam, say, or a problem with the toner—the machines would flash a cryptic code number, which would require more flipping through the cards to find the corresponding explanation.

In many instances, users would encounter some obstacle, not be able to find out how to resolve it, and simply abandon the machine in mid-procedure. The next user to come along, unaware of the previous problem, would assume the machine was broken and call a repair person.

We had to make radical changes in copier design, but it was difficult to sell that message within the company. The idea that there might be serious usability problems with our machines met with resistance in the Xerox development organization that designs our copiers. After all, they had tested their designs against all the traditional "human factors" criteria. There was a tendency to assume that any problems with the machines must be the users' fault.

When researchers from PARC began to study the problem, we discovered that the human-factors tests used by the development group didn't accurately reflect how people actually used the machines. So, a PARC anthropologist set up a video camera overlooking one of our new copiers in use at PARC, then had pairs of researchers (including some leading computer scientists) use the machine to do their own copying. The result was dramatic footage of some very smart people, anything but idiots, becoming increasingly frustrated and angry as they tried and failed to figure out how to get the machine to do what they wanted it to do.

The videos proved crucial in convincing the doubters that the company had a serious problem. Even more important, they helped us define what the real problem was. The videos demonstrated that when people use technology like a copier, they construct interpretations of it. In effect, they have a conversation with the machine much as two people have a conversation with each other. But our traditional idiot-proof design provided few cues to help the user interpret what was going on.

We proposed an alternative approach to design. Instead of trying to eliminate "trouble," we acknowledged that it was inevitable. So the copier's design should help users "manage" trouble—just as people manage and recover from misunderstandings during a conversation. This meant keeping the machine as transparent as possible by making it easy for the user to find out what is going on and to discover immediately what to do when something goes wrong.

Xerox's most recent copier families—the 10 and 50 series—reflect this new design principle. Gone are the flip cards of earlier machines. Instead, we include enough computing power in the machines to provide customized instructions on the display panel linked to particular procedures or functions. The information the user receives is immediately put in the context of the task he or she is trying to perform. The new design also incorporates ideas from PARC's research on graphical user interfaces for computers. When something goes wrong, the display panel immediately shows a picture of the machine that visually indicates where the problem is and how to resolve it.

The results of these changes have been dramatic. Where it once took 28 minutes on average to clear a paper jam, it takes 20 seconds with the new design. And because such breakdowns are easier to fix, customers are more tolerant of them when they occur.

But these changes are only the beginning. Once copiers become computing devices, they also become sensors that collect information about their own performance that can be used to improve service and product design. For example, Xerox recently introduced a new standard feature on our high-end copiers known as "remote interactive communication" or RIC. RIC is an expert system inside the copier that monitors the information technology controlling the machine and, using some artificial-intelligence techniques, predicts when the machine will next break down. Once RIC predicts a breakdown will occur, it automatically places a call to a branch office and downloads its prediction, along with its reasoning. A computer at the branch office does some further analysis and schedules a repair person to visit the site *before* the expected time of failure.

For the customer, RIC means never having to see the machine fail. For Xerox, it means not only providing better service but also having a new way to "listen" to our customer. As RIC collects information on the performance of our copiers—in real-world business environments, year in and year out—we will eventually be able to use that information to guide how we design future generations of copiers.

RIC is one example of how information technology invisible to the user is transforming the copier. But the ultimate conclusion of this technological transformation is the disappearance of the copier as a stand-alone device. Recently, Xerox introduced its most versatile office machine ever—a product that replaces traditional light-lens copying techniques with "digital copying," where documents are electronically scanned to create an image stored in a computer, then printed out whenever needed. In the future, digital copiers will allow the user to scan a document at one site and print it out somewhere else—much like a fax. And once it scans a document, a copier will be able to store, edit, or enhance the document—like a computer file—before printing it. When this happens, the traditional distinction between the copier and other office devices like computers, printers, and fax machines will disappear—leaving a flexible, multifunctional device able to serve a variety of user needs.

What is happening to the copier will eventually happen to all office devices. As computing power becomes ubiquitous—incorporated not only in copiers but also in filing cabinets, desktops, white boards, even electronic "Post-it" notes—it will become more and more invisible, a taken-for-granted part of any work environment, much as books, reports, or other documents are today. What's more, increased computing power will make possible new uses of information technology that are far more flexible than current systems. In effect, technology will become so flexible that users will be able to customize it ever-more precisely to meet their particular needs—a process that might be termed "mass customization."

We are already beginning to see this development in software design. Increased computing power is making possible new approaches to writing software such as "object-oriented programming" (developed at PARC in the 1970s). This technique makes it easier for users to perform customizing tasks that previously required a trained programmer and allows them to adapt and redesign information systems as their needs change. From a purely technical perspective, object-oriented programming may be less efficient than traditional programming techniques. But the flexibility it makes possible is far more suited to the needs of constantly evolving organizations.

Indeed, at some point in the not-too-distant future—certainly within the next decade—information technology will become a kind of generic entity, almost like clay. And the "product" will not exist until it enters a specific situation, where vendor and customer will mold it to the work practices of the customer organization. When that happens, information technology as a distinct category of products will become invisible. It will dissolve into the work itself. And companies like ours might sell not products but rather the expertise to help users define their needs and create the products best suited to them. Our product will be our customers' learning.

Harvesting Local Innovation

The trend toward ubiquitous computing and mass customization is made possible by technology. The emphasis, however, is not on the technology itself but on the work practices it supports. In the future, organizations won't have to shape how they work to fit the narrow confines of an inflexible technology. Rather, they can begin to design information systems to support the way people really work.

That's why some of the most important research at PARC in the past decade has been done by anthropologists. PARC anthropologists have studied occupations and work practices throughout the company—clerks in an accounts-payable office who issue checks to suppliers, technical representatives who repair copying machines, designers who develop new products, even novice users of Xerox's copiers. This research has produced fundamental insights into the nature of innovation, organizational learning, and good product design.

We got involved in the anthropology of work for a good business reason. We figured that before we went ahead and applied technology to work, we had better have a clear understanding of exactly how people do their jobs. Most people assume—we did too, at first—that the formal procedures defining a job or the explicit structure of an organizational chart accurately describe what employees do, especially in highly routinized occupations. But when PARC anthropologist Lucy Suchman began studying Xerox accounting clerks in 1979, she uncovered an unexpected and intriguing contradiction.

When Suchman asked the clerks how they did their jobs, their descriptions corresponded more or less to the formal procedures of the job manual. But when she observed them at work, she discovered that the clerks weren't really following those procedures at all. Instead, they relied on a rich variety of informal practices that weren't in any manual but turned out to be crucial to getting the work done. In fact, the clerks were constantly improvising, inventing new methods to deal with unexpected difficulties and to solve immediate problems. Without being aware of it, they were far more innovative and creative than anybody who heard them describe their "routine" jobs ever would have thought.

Suchman concluded that formal office procedures have almost nothing to do with how people do their jobs. People use procedures to understand the goals of a particular job—for example, what kind of information a particular file has to contain in order for a bill to be paid—not to identify the steps to take in order to get from here to there. But in order to reach that goal—actually collecting and verifying the information and making sure the bill is paid—people constantly invent new work practices to cope with the unforeseen contingencies of the moment. These informal activities remain mostly invisible since they do not fall within the normal, specified procedures that employees are expected to follow or managers expect to see. But these "workarounds" enable an all-important flexibility that allows organizations to cope with the unexpected, as well as to profit from experience and to change.

If local innovation is as important and pervasive as we suspect, then big companies have the potential to be remarkably innovative—*if* they can somehow capture this innovation and learn from it. Unfortunately, it is the rare company that understands the importance of informal improvisation—let alone respects it as a legitimate business activity. In most cases, ideas generated by employees in the course of their work are lost to the organization as a whole. An individual might use them to make his or her job easier and perhaps even share them informally with a small group of colleagues. But such informal insights about work rarely spread beyond the local work group. And because most information systems are based on the formal procedures of work, not the informal practices crucial to getting it done, they often tend to make things worse rather than better. As a result, this important source of organizational learning is either ignored or suppressed.

At PARC, we are trying to design new uses of technology that leverage the incremental innovation coming from within the entire company. We want to create work environments where people can legitimately improvise, and where those improvisations can be captured and made part of the organization's collective knowledge base.

One way is to provide people with easy-to-use programming tools so they can customize the information systems and computer applications that they work with. To take a small example, my assistant is continually discovering new ways to improve the work systems in our office. She has more ideas for perfecting, say, our electronic calendar system than any researcher does. After all, she uses it every day and frequently bumps up against its limitations. So instead of designing a new and better calendar system, we created a programming language known as CUSP (for "customized user-system program") that allows users to modify the system themselves.

We've taken another small step in this direction at EuroPARC, our European research lab in Cambridge, England. Researchers there have invented an even more advanced software system known as "Buttons"—bits of computer code structured and packaged so that even people without a lot of training in computers can modify them. With Buttons, secretaries, clerks, technicians, and others can create their own software applications, send them to colleagues throughout the corporation over our electronic mail network, and adapt any Buttons they receive from others to their own needs. Through the use of such tools, we are translating local innovation into software that can be easily disseminated and used by all.

New technologies can also serve as powerful aids for organizational learning. For example, in 1984 Xerox's service organization asked us to research ways to improve the effectiveness of their training programs. Training the company's 14,500 service technicians who repair copying machines is extremely costly and time-consuming. What's more, the time it takes to train the service work force on a new technology is key to how fast the company can launch new products.

The service organization was hoping we could make traditional classroom training happen faster, perhaps by creating some kind of expert system. But based on our evolving theory of work and innovation, we decided to take another approach. We sent out a former service technician, who had since gone on to do graduate work in anthropology, to find out how reps actually do their jobs—not what they or their managers say they do but what they really do and how they learn the skills that they actually use. He took the company training program, actually worked on repair jobs in the field, and interviewed tech-reps about their jobs. He concluded that the reps learn the most not from formal training courses but out in the field— by working on real problems and discussing them informally with colleagues. Indeed, the stories tech-reps tell each other—around the coffee pot, in the lunchroom, or while working together on a particularly difficult problem—are crucial to continuous learning.

In a sense, these stories are the real "expert systems" used by tech-reps on the job. They are a storehouse of past problems and diagnoses, a template for constructing a theory about the current problem, and the basis for making an educated stab at a solution. By creating such stories and constantly refining them through conversation with each other, tech-reps are creating a powerful "organizational memory" that is a valuable resource for the company.

As a result of this research, we are rethinking the design of tech-rep training—and the tech-rep job itself—in terms of lifelong learning. How might a company support and leverage the storytelling that is crucial to building the expertise not only of individual tech-reps but also of the entire tech-rep community? And is there any way to link that expertise to other groups in the company who would benefit from it—for example, the designers who are creating the future generations of our systems?

One possibility is to create advanced multimedia information systems that would make it easier for reps and other employees to plug in to this collective social mind. Such a system might allow the reps to pass around annotated videoclips of useful stories, much like scientists distribute their scientific papers, to sites all over the world. By commenting on each other's experiences, reps could refine and disseminate new knowledge. This distributed collective memory, containing all the informal expertise and lore of the occupation, could help tech-reps—and the company—improve their capacity to learn from successes and failures.

Coproducing Innovation

Our approach to the issue of tech-rep training is a good example of what we mean by "pioneering" research. We started with a real business problem, recognized by everyone, then reframed the problem to come up with solutions that no one had considered before. But this raises another challenge of pioneering research: How to communicate fresh insights about familiar problems so that others can grasp their significance?

The traditional approach to communicating new innovations—a process that usually goes by the name of "technology transfer"—is to treat it as a simple problem of transferring information. Research has to pour new knowledge into people's heads like water from a pitcher into a glass. That kind of communication might work for incremental innovations. But when it comes to pioneering research that fundamentally redefines a technology, product, work process, or business problem, this approach doesn't work.

It's never enough to just *tell* people about some new insight. Rather, you have to get them to experience it in a way that evokes its power and possibility. Instead of pouring knowledge into people's heads, you need to help them grind a new set of eyeglasses so they can see the world in a new way. That involves challenging the implicit assumptions that have shaped the way people in an organization have historically looked at things. It also requires creating new communication techniques that actually get people to experience the implications of a new innovation.

To get an idea of this process, consider the strategic implications of an innovation such as digital copying for a company like Xerox. Xerox owes its existence to a particular technology—light-lens xerography. That tradition has shaped how the company conceives of products, markets, and customer needs, often in ways that are not so easy to identify. But digital copying renders many of those assumptions obsolete. Therefore, making these assumptions explicit and analyzing their limitations is an essential strategic task.

Until recently, most people at Xerox thought of information technology mainly as a way to make traditional copiers cheaper and better. They didn't realize that digital copying would transform the business with broad implications not just for copiers but also for office information systems in general. Working with the Xerox corporate strategy office, we've tried to find a way to open up the corporate imagination—to get people to move beyond the standard ways they thought about copiers.

One approach we took a couple of years ago was to create a video for top management that we called the "unfinished document." In the video, researchers at PARC who knew the technology extremely well discussed the potential of digital copying to transform people's work. But they didn't just talk about it; they actually acted it out in skits. They created mock-ups of the technology and then simulated how it might affect different work activities. They attempted to portray not just the technology but also the technology "in use."

We thought of the unfinished document as a "conceptual envisioning experiment"—an attempt to imagine how a technology might be used before we started building it. We showed the video to some top corporate officers to get their intuitional juices flowing. The document was "unfinished" in the sense that the whole point of the exercise was to get the viewers to complete the video by suggesting their own ideas for how they might use the new technology and what these new uses might mean for the business. In the process, they weren't just learning about a new technology; they were creating a new mental model of the business.

Senior management is an important partner for research, but our experiments at coproduction aren't limited to the top. We are also involved in initiatives to get managers far down in the organization to reflect on the obstacles blocking innovation in the Xerox culture. For example, one project takes as its starting point the familiar fact that the best innovations are often the product of "renegades" on the periphery of the company. PARC researchers are part of a company group that is trying to understand why this is so often the case. We are studying some of the company's most adventuresome product-development programs to learn how the larger Xerox organization can sometimes obstruct a new product or work process. By learning how the corporation rejects certain ideas, we hope to uncover those features of the corporate culture that need to change.

Such efforts are the beginning of what we hope will become an ongoing dialogue in the company about Xerox's organizational practice. By challenging the background assumptions that traditionally stifle innovation, we hope to create an environment where the creativity of talented people can flourish and "pull" new ideas into the business.

Innovating with the Customer

Finally, research's ultimate partner in coproduction is the customer. The logical end point of all the activities I have described is for corporate research to move outside the company and work with customers to coproduce the technology and work systems they will need in the future.

It is important to distinguish this activity from conventional market research. Most market research assumes either that a particular product already exists or that customers already know what they need. At PARC, we are focusing on systems that do not yet exist and on needs that are not yet clearly defined. We want to help customers become aware of their latent needs, then customize systems to meet them. Put another way, we are trying to prototype a need or use before we prototype a system.

One step in this direction is an initiative of Xerox's Corporate Research Group (of which PARC is a part) known as the Express project. Express is an experiment in product-delivery management designed to commercialize PARC technologies more rapidly by directly involving customers in the innovation process. The project brings together in a single organization based at PARC a small team of Xerox researchers, engineers, and marketers with employees from one of our customers—Syntex, a Palo Alto-based pharmaceutical company.

Syntex's more than 1,000 researchers do R&D on new drugs up for approval by the Food and Drug Administration. The Express team is exploring ways to use core technologies developed at PARC to help the pharmaceutical company manage the more than 300,000 "case report" forms it collects each year. (The forms report on tests of new drugs on human volunteers.) Syntex employees have spent time at PARC learning our technologies-in-progress. Similarly, the Xerox members of the team have intensively studied Syntex's work processes—much as PARC anthropologists have studied work inside our own company.

Once the project team defined the pharmaceutical company's key business needs and the PARC technologies that could be used to meet them, programmers from both companies worked together to create some prototypes. One new system, for example, is known as the Forms Receptionist. It combines technologies for document interchange and translation, document recognition, and intelligent scanning to scan, sort, file, and distribute Syntex's case reports. For Syntex, the new system solves an important business problem. For Xerox, it is the prototype of a product that we eventually hope to offer to the entire pharmaceutical industry.

We are also treating Express as a case study in coproduction, worth studying in its own right. The Express team has videotaped all the interactions between Xerox and Syntex employees and developed a computerized index to guide it through this visual database. And a second research team is doing an in-depth study of the entire Xerox-Syntex collaboration. By studying the project, we hope to learn valuable lessons about coproduction.

For example, one of the most interesting lessons we've learned from the Express project so far is just how long it takes to create a shared understanding among the members of such product teams—a common language, sense of purpose, and definition of goals. This is similar to the experience of many interfunctional teams that end up reproducing inside the team the same conflicting perspectives the teams were designed to overcome in the first place. We believe the persistence of such misunderstandings may be a serious drag on product development.

Thus a critical task for the future is to explore how information technology might be used to accelerate the creation of mutual understandings within work groups. The end point of this process would be to build what might be called an "envisioning laboratory"—a powerful computer environment where Xerox customers would have access to advanced programming tools for quickly modeling and envisioning the consequences of new systems. Working with Xerox's development and marketing organizations, customers could try out new system configurations, reflect on the appropriateness of the systems for their business, and progressively refine and tailor them to match their business needs. Such an environment would be a new kind of technological medium. Its purpose would be to create evocative simulations of new systems and new products before actually building them.

The envisioning laboratory does not yet exist. Still, it is not so farfetched to imagine a point in the near future where major corporations will have research centers with the technological capability of, say, a multimedia computer-animation studio like Lucasfilm. Using state-of-the-art animation techniques, such a laboratory could create elaborate simulations of new products and use them to explore the implications of those products on a customer's work organization. Prototypes that today take years to create could be roughed out in a matter of weeks or days.

When this happens, phrases like "continuous innovation" and the "customer-driven" company will take on new meaning. And the transformation of corporate research—and the corporation as a whole—will be complete.

2

The Designer Organization: Italy's GFT Goes Global

Robert Howard

Not so long ago, most managers had a clear picture of what they meant by a global company. At the center, a corporate headquarters (and for the majority of U.S. managers, that meant a U.S. headquarters) was the source of investment capital, new products, and managerial decisions. On the periphery, foreign subsidiaries used these resources to develop local markets. Innovation and learning moved from the center to the periphery. If all went well, profits flowed back.

In the last decade, however, new realities of global competition have thoroughly scrambled that traditional picture. It's not just that more and more U.S. managers are now on the periphery of some foreign company's global business. As manufacturing, product design, and other key managerial functions migrate to a company's key markets around the world, the very meaning of "center" and "periphery"—and the understanding of what constitutes the proper relationship between them—have come into question. And while it's one thing to urge companies to be simultaneously "global" and "local," it's quite another to develop the right kind of capabilities and the right organization to manage this delicate balancing act.

Consider the example of the Italian apparel manufacturer Gruppo GFT. Based in Turin, Italy, GFT is the world's largest manufacturer of designer clothing. It competes at the highest end of the apparel business—"ready-to-wear" designer collections, one step below made-to-order haute couture. GFT is the company behind such well-known European designer labels as Giorgio Armani, Emanuel Ungaro, and Valentino. The company makes some of the highest quality and most expensive clothes in the world: men's suits and women's outfits that sell for as much as $1,200.

Until recently, GFT was a relatively small and primarily Italian company. But in the 1980s, it rode the wave of global interest in European fashion to become a billion dollar business, with about 60% of its sales outside Italy and 26% in the United States alone. GFT brought the whole "Made in Italy" (synonymous with high quality and European design) fashion craze to the world. The company almost single-handedly created the U.S. market for Italian designer apparel. Its U.S. sales grew from $7 million in 1980 to $304 million in 1989. Today GFT's 10,000 employees in the 45 small companies and 18 manufacturing plants under the Gruppo GFT umbrella make, distribute, and market roughly 60 designer and branded collections in 70 countries around the world.

But in the process of becoming a global company, GFT's managers have discovered that being global means something very different from what they had originally thought. They are struggling with a set of paradoxes that more and more companies are facing.

For GFT, globalization is not about standardization; it's about a quantum increase in complexity. The more the company has penetrated global markets, the more sustaining its growth depends on responding to myriad local differences in its key markets around the world. In the words of GFT chairman Marco Rivetti, "to be global means to recognize difference and be flexible enough to adapt to it."

But to achieve this flexibility requires turning the organization inside out. In a sense, the periphery has to become the center—or, at least, the center of top management's attention. Adapting to local differences requires a far more multifaceted organizational structure, one in which innovation occurs at the periphery as well as the center and where learning flows in many different directions. The centerpiece of GFT's global business strategy is to become an "insider" in each of its major markets.

Becoming an insider, however, has profound organizational implications. Put simply, GFT has had to reinvent the way it does business—how it defines its customers, how it develops and manufactures products, how it handles marketing and distribution. What's more, this reinvention is not a onetime event but an ongoing process. In effect, GFT is trying to create a "designer organization," able to adjust and adapt continuously to differences among markets and changes within markets—much as GFT's designer collections change from year to year. The chief role of GFT's global managers is to manage the continuous redesign of the company.

What follows then is the story of one company's efforts to create an organization adequate to the new complexities of global competition. Think of it as a "Tale of Three Cities":

Turin, Italy is GFT's corporate center and source of the company's core expertise, but it is also the place where the traditional mentality of company managers needs to change the most.

New York, New York is home of GFT USA Corp., the company's largest foreign subsidiary in its most dynamic market, and the key expression of GFT's insider strategy.

New Bedford, Massachusetts is home of the Riverside Manufacturing Co., a small outpost of GFT's global production system and a vivid example of the company's ever-more complex flows of innovation and learning.

Three cities, three snapshots of the new global corporation.

Turin: "We Have to Reinvent Everything from Scratch"

There is no missing GFT's corporate headquarters in Turin, Italy. In 1984, the company hired the Italian postmodern architect Aldo Rossi to renovate its main offices. The result looks like some incongruous Egyptian temple set down in a traditional industrial neighborhood— more museum of contemporary art than headquarters of a global apparel manufacturer. And yet, Rossi's design is an apt symbol, both of GFT's transformation into a global company in the 1980s and the further transformation the company faces today.

During the period following the Second World War, GFT was the first European apparel company to introduce mass-production techniques. In a sense, the company spearheaded the "Americanization" of the Italian apparel industry by importing production technology and know-how—in the person of Italian-American apparel workers— from the United States.

Throughout the 1950s and 1960s, GFT established itself as a highly efficient producer of standardized men's clothing. Its production system was rigid, but the company's capacity to produce in high volume meant that unit costs were low. This allowed GFT to sell to the emerging mass market for men's clothing in Italy and even to develop a minor business making low-cost, private-label suits for U.S. retailers such as Marshall Field and B. Altman.

By the early 1970s, however, the company's position as a low-cost producer was becoming harder and harder to sustain. One factor was the massive labor unrest that occurred throughout Italy in the late 1960s. Worker militancy led to new, more restrictive labor laws—in particular, the government's decision in 1975 to index workers' wages to the inflation rate. Companies like GFT were saddled with a rigid cost structure that pushed up wages, cut into profits, and eliminated the traditional wage advantage of Italian apparel manufacturers in foreign markets.

For Marco Rivetti, who headed GFT's womenswear division in the 1970s and succeeded his father as chairman of the company in 1986, the solution was to move up-market, to broaden the company's product line by entering niches at the high end of the apparel industry. Competing in these niches would not only be more profitable but it would also allow the company to respond to the growing interest in more expensive and more stylish clothing.

Rivetti's signal innovation for accomplishing this shift was to forge close alliances with top fashion designers in Italy and France. Starting in the 1970s, GFT began signing licensing agreements with well-known fashion designers such as Ungaro, Armani, and Valentino. By the late 1980s, the company had agreements with ten European designers.

Its relationships with the designers changed GFT in a number of ways. Most important, it transformed GFT's production process. Before, the company had been "production-driven," its business defined by standardized products that could be manufactured in high volume and at low cost. Now GFT became "designer-driven," combining the company's traditional focus on efficiency with a new emphasis on flexibility and extremely high quality.

One sign of flexibility is the company's expansion into womenswear and sportswear, which together now make up roughly 37% of sales. But even in its traditional menswear business, GFT has had to become a far more flexible manufacturer. The company's Italian plants can make 500 different models of a man's suit jacket. As for quality, at the company's menswear plant in Settimo, outside Turin (where Armani jackets are made for sale around the world), workers examine the entire length of every bolt of fabric for flaws before sending the fabric into production. And at the end of the production process, before the jackets go out, a single individual tries on about 300 jackets every day to make sure the fit is right.

During GFT's designer-driven era, being global meant functioning like a traditional multinational. Innovations in product development, manufacturing, and business strategy took place in Turin and were then exported around the world. Even as foreign markets grew in importance, the business was run and controlled from Turin. GFT manufactured global products and shipped them worldwide: the Armani jacket sold in Milan was identical to the one sold in New York or Tokyo. And the company grew on the wave of global interest in European fashion.

By the end of the 1980s, however, Rivetti and other top GFT managers could see that this formula no longer made sense. Three related developments are forcing GFT to reinvent itself once again.

The first is globalization itself. As more and more of the company's business is outside Italy, GFT has discovered that the collections of even its "international" designers don't travel well without some adaptation. Pieces that are best-sellers in Italy often do poorly in the United States. Sizes, fabrics, and colors that work in Germany or England aren't right for Florida or California. So GFT's "global" products have to be adapted to meet the very different needs and expectations of different local markets. In the words of one GFT manager in Turin, "the company may be global, but the consumer is not."

At the same time, the market for designer clothing has become a lot tougher. It's not just that recession in key markets like the United States is dampening demand (weak demand and the low buying power of the dollar were major factors in depressing GFT's 1990 net earnings by about 75%). Paradoxically, the very success of designer fashions in the 1980s means that consumers have become more educated and more discerning about fashion.

For example, one result of the great success of designer labels in the 1970s and 1980s was that designer collections expanded to include nearly everything—not just men's suits but also shirts, socks, handkerchiefs, even umbrellas and sunglasses. Just having the designer name was enough to sell the entire line. Similarly, the enormous success of some designers like Armani served to carry other less well-known designers in GFT's stable. The designer-label phenomenon was enough to get them into the stores and into customer's hands.

But the mere fact of a designer label or knowledge that a product is "Made in Italy" isn't enough anymore. Customers still want high quality, but they want it at a lower price. The emphasis is on "value" and a much tighter trade-off between quality and cost. Every item of a particular designer collection has to be competitive on price or customers will go elsewhere. Similarly, each designer has to be profitable or retailers won't carry them in the stores. This is pushing GFT to simplify its collections and to make strategic bets among its various collections and designers, keeping some and jettisoning others depending on how closely they track the market.

Finally, while there will always be an international elite willing and able to buy whatever the company's designer partners offer, this market is far from adequate to sustain GFT's growth. This is forcing GFT along with other designer apparel companies to develop a new kind of market in so-called "diffusion" lines: designer collections one notch down-market from ready-to-wear lines.

Diffusion lines are where the future growth in the designer apparel industry is likely to be. For instance, in 1989 U.S. designer Donna Karan introduced a women's diffusion collection called DKNY. Annual sales are expected to reach $135 million by the end of this year.

Competing in the diffusion market means a 180-degree turn in managerial competencies. Take the example of the GFT product manager, a position that in the designer-driven era of the 1970s and 1980s became a symbol of the company's transformation. The product manager was the liaison between GFT and the fashion designer. In effect, his or her job was to translate the vision of the designer as perfectly as possible into products, no matter what the cost. A whole generation of product managers learned how to make themselves, and GFT as a whole, the agents of a designer's vision. They became experts at finding the best fabrics or learning how to execute the most complicated trim. Price was not really an issue; image and quality were.

Now product managers have to be more sensitive to the trade-off between quality and cost. They have to choose fabrics and processes with price in mind as well as quality. Even more important, they have to know the market, precisely what kind of customer a particular collection is aimed at and exactly what that customer wants. And they have to play a more active role in their relationships with the fashion designers, educating them to the needs of the market. "The product manager has to become a marketing manager," says Rivetti. Not all of GFT's traditional product managers can make the transition. In the past two years, for example, the company has replaced many of the product managers in its womenswear division.

Getting closer to the market also means turning the organization inside out—specifically, by giving GFT's subsidiaries in key markets around the world enough autonomy so that they can mold themselves to the particularities of their local market, what the company calls becoming an "insider." Where the alliances with fashion designers were the change agents of GFT's previous transformation, it is these insider subsidiaries in markets such as the United States that are the new agents of change.

Marco Rivetti puts it this way: "Traditionally, we delegated the creativity to the designers and the relationship with the customer to the retailer. Now, *we* have to do both. We have to reinvent everything from scratch."

New York: "An Italian Interpretation of an American Company"

From the 35th floor of a Fifth Avenue office building, in a space that previously housed the trading room of financier Ivan Boesky, the managers of GFT USA Corp. are shaping GFT's business to the specific needs of the U.S. market. GFT USA is Gruppo GFT's largest foreign subsidiary and an expression of the company's attempt to balance the local autonomy of an "insider" with a global company's coordination.

Until quite recently, GFT saw the United States primarily as an export market. Its U.S. business, including subsidiaries like GFT USA, was run directly from Turin—first by an export-import department, then by an international division that the company established in 1985.

But the need to get closer to local markets caused GFT to create a new, more decentralized organizational structure in 1989. Today GFT is a financial holding company made up of six autonomous operating divisions. Two are based on key global markets: North America (the United States, Canada, and Mexico) and the Far East. The remaining four operating divisions are based on GFT's three main product lines— menswear, womenswear, sportswear—and on the company's traditional but increasingly less important fabrics business. The four product divisions focus on the European market.

GFT USA is the flagship company of the North American division. The company is responsible for meeting an annual profit objective set in collaboration with top managers in Turin. But within the boundaries of that financial objective, GFT USA managers are free to run their business as they see fit.

GFT USA's own organization repeats this decentralized structure. The company itself is a holding company consisting of six small companies, four based on product lines and one each for GFT's U.S.-based manufacturing and distribution operations. Another six Gruppo GFT companies also do business in the United States. Although not formally a part of GFT USA, they coordinate their efforts with the company and sometimes share the costs of joint projects.

The purpose of this decentralized structure is to give GFT USA the freedom to respond quickly to the distinctive features of the U.S. market. That's important because the United States represents such a large part of Gruppo GFT's business, but there is another reason as well. Frequently, trends in the United States prefigure developments that eventually occur in other markets around the world. Therefore, GFT USA is a kind of lab for business innovations that will prove important to the future of the group as a whole.

One such innovation launched by GFT USA this year is a new women's diffusion line, Emanuel by Emanuel Ungaro. The Ungaro design house is based in Paris, but the Emanuel collection was developed at GFT USA to meet the specific needs of U.S. working women.

In the late 1980s, GFT had only one women's diffusion collection on sale in the United States, another Ungaro line known as Ungaro Ter. Ungaro Ter was a reinterpretation of a traditional European designer product, offered at a more affordable price. But interviews with customers and salespeople in stores and analysis of weekly sales data convinced GFT USA's strategic marketing department that the collection was not really meeting the needs of the U.S. market. The design was wrong: Ungaro Ter outfits were fine for work, but customers wanted clothing they could also wear for evenings out or on weekends, or use in more than one season. The mix of the collection was wrong: newly popular items such as trousers and jumpsuits were missing. The range of fabrics was wrong—lacking new materials like stretch woolens and stretch corduroys. Most important, the fit was wrong—too restrictive, too "French" for American tastes.

Based on these findings, in the spring of 1990 GFT USA developed the concept for a new Ungaro diffusion line that would respond to these emerging customer needs and in which every item would sell for under $500. But first the company had to educate the Ungaro design house about the new trends in the American market. GFT USA managers traveled to Paris to present their market research and sell the new concept. And Emanuel Ungaro traveled incognito into some retail stores in California to talk with customers and salespeople.

Once sold on the concept, the Ungaro design house developed sketches for the new Emanuel line. But the all-important next step of developing prototypes for the collection took place not in Paris but in New York at GFT USA's own sample rooms. There, GFT USA technical designers modified and adapted the ideas embodied in the sketches, often in close collaboration with their counterparts in Paris. And GFT USA merchandisers packaged the collection, selecting the right mix of products and the right fabrics and colors. By the end of 1990, the prototypes were completed and Emanuel Ungaro had signed off on the final versions. By February 1991, less than a year after GFT USA came up with the Emanuel concept, retailers were placing their orders for a July delivery of the fall 1991 Emanuel collection.

With the Emanuel diffusion line, GFT USA is turning the designer label into a kind of brand. Instead of the designer developing a product concept and GFT USA having to adapt it to the U.S. market, GFT USA's knowledge of the market gives birth to the product concept in the first place.

This customer-driven design process has direct implications for how the Emanuel collection has been structured. In the past, GFT delivered womenswear collections to the retailer four times a year. With Emanuel, the company will deliver a different segment from GFT USA's New Jersey warehouse every month, thus constantly refreshing the collection in the stores and offering customers a steady stream of new products. To make more frequent distribution possible, GFT USA has organized the design and production of the collection around segments that emphasize different combinations of colors, fabric weights, and functions (for instance, clothes for work, for leisure, and for evenings out) depending on the season.

The emphasis on getting closer to the customer embodied in the Emanuel collection also means getting closer to the retailer. Not long ago, what happened after a department store or specialty boutique ordered a collection—in particular, whether the product ever got into the hands of the consumer—was not GFT's concern. Now when buyers come to GFT USA's Fifth Avenue showrooms, in addition to looking at the clothes, they peruse a "boutique catalogue" with special accessories for creating the unique store environment accompanying each product line. For Emanuel and all its other collections, GFT USA designs the shop, the fixtures, the furniture, even the plants. The company also trains the retailer's salespeople in how to sell the product.

The new Emanuel line suggests how GFT's insider strategy is transforming its traditional European designer product. Another insider initiative is taking GFT beyond its traditional focus on European design. The company is developing relationships with local fashion designers whose work captures the style or customer wants of a particular region or country. Here, again, the key initiatives are taking place at GFT USA. GFT's two most recent designer partners are Andrew Fezza and Joseph Abboud, American designers with a distinctive

American style that nevertheless fits into GFT's global image of high quality and good design. The bulk of the Fezza and Abboud menswear collections are not "Made in Italy" but at GFT manufacturing plants in Mexico and the United States.

The agreements with Fezza and Abboud represent an organizational innovation as well as a product innovation. Instead of the traditional licensing arrangement, both agreements are actual joint ventures, subsidiaries of GFT USA in which GFT holds the majority interest. This is yet another way to link designers more closely to the market and to make sure their offerings are responsive to the business environment.

Becoming an insider, however, doesn't mean that GFT USA isn't an integral part of GFT's global business. Indeed, the more "local" GFT USA has become, the more "global" it has become as well.

First, the company's local initiatives are intimately shaped by GFT's global strategy. Three years ago, GFT introduced a strategic planning system, the company's first. For a company as decentralized as GFT, establishing a strategic direction that coordinates the work of its 45 companies and 18 manufacturing plants around the world is crucial. But given the volatile nature of the fashion business and large differences among markets, any planning system has to be simple and extremely flexible.

GFT's planning process is highly collaborative, reaching down to each individual business unit. For example, each subsidiary of GFT USA develops its own strategy: analyzing the structure and price of its collections, developing a distribution strategy for keeping major accounts and reaching out to new ones, defining the unit's needs for investments, functional support, and technical expertise from corporate. Managers in New York work with executives in Turin to dovetail the plans with GFT's global priorities. For instance, too many new product launches at any one time can strain the group's financial resources. Decisions made in New York can affect GFT's global relations with its designer partners. In the end, the decision about whether a particular initiative moves forward is made in Turin but with considerable input from New York.

Another way GFT USA is linked to GFT's global efforts is through people. The company's CEO, Leopoldo Borzino, is also head of GFT's North American division. Previously, he served as director of GFT's international and export divisions, which originally opened the American market for GFT in the mid-1970s.

Or take the example of Marina Mira d'Ercole, director of strategic planning at GFT USA. Her recent career trajectory has been an exercise in managerial "technology transfer." Before coming to New York in 1989, she helped set up GFT's new strategic planning system. Her chief responsibility during her two-year stay in the United States has been to teach that system to managers at GFT USA. Later this year, she will return to Italy to reverse the learning process, most likely by taking a leadership position in a unit linked closely with the U.S. business, so as to spread her knowledge of the U.S. market and GFT's U.S. operations to managers in Turin.

Finally, GFT USA is integrated into the business activities of the group as a whole through the expectation that business innovations developed in the United States will eventually travel elsewhere. The development of the Emanuel line has important lessons for how GFT interacts with its designer partners and with retailers all over the world, not just in the United States. Similarly, Joseph Abboud may be an American designer, but GFT has made the strategic bet that he is good enough to serve global markets. The company is distributing Joseph Abboud clothing in Italy, which means adapting the designer's collections to local tastes and expectations.

All these examples suggest that the flows of innovation and learning within GFT are far more complex today than they were as recently as five years ago. So is the company's corporate identity. Is GFT USA an American company or an Italian one? Neither, says Leopoldo Borzino: "We are an Italian interpretation of an American company."

New Bedford: From "Made in Italy" to "Made in GFT"

The entrance to the Riverside Manufacturing Co. is an easy-to-miss wooden door at the end of a long redbrick mill building in New Bedford, Massachusetts. "We are like a small outpost on the frontier of the Roman Empire," says Umberto Caccina, the 33-year GFT veteran who has spent the last four years integrating Riverside into GFT's global operations. Ironically, Riverside's location on the far periphery of the GFT empire makes it a central example of the company's more intricate business relationships.

Riverside is GFT USA's only manufacturing subsidiary. Its 250 employees produce 70,000 pieces of clothing a year for one brand and three designer labels—including the Joseph Abboud men's ready-to-wear and diffusion collections.

GFT bought Riverside in 1987 to establish a small production base in the United States and to enhance GFT's insider strategy to get closer to the U.S. market. For more than 20 years, GFT has been exporting its distinctive manufacturing know-how, which combines the efficiencies of mass production with high levels of flexibility and quality. The goal: to create a worldwide production network that will make GFT a guarantor of high quality and good value—no matter where the company's products are made. GFT corporate managers in Turin describe the goal this way: "We have to move from 'Made in Italy' to 'Made in GFT.'" Riverside suggests how they are doing it.

At first, most of the learning that took place at Riverside was "one-way"—from Turin to New Bedford. Before GFT came on the scene, Riverside worked as a contract shop making men's and women's tailored jackets for U.S. companies such as Hartmarx and Liz Claiborne. As a contractor, the company had no design capability and no real manufacturing system. Procedures changed with each customer. The haphazard flow of work meant frequent layoffs and high employee turnover.

Ironically, the shop's history as a contractor was an important reason why GFT chose Riverside when it was shopping for an American acquisition. Managers reasoned that because Riverside workers had to learn the procedures and work systems of the company's many different customers, they would have the flexibility to learn the GFT system quickly as well.

Over the next year and a half, GFT invested $1.5 million in technology and training at Riverside, changing every machine and organizational process in the company's New Bedford plant. For example, Riverside had never had an information system. Now the company has an automated marking and production planning system linked to GFT USA headquarters in New York. Based on retailer orders, the system schedules production—even to the point of determining the specific retail customer for each piece of clothing before it is made. This allows workers to customize each jacket by sewing in the retailer's label in addition to that of the designer.

But the more important changes at Riverside have been organizational. To maximize efficiency and quality, GFT has a very specific methodology for every single operation in its factories. Often, these techniques are surprisingly simple. One small example: the cut fabric swatches used for the inside of jacket pockets are kept in small cardboard folders to ensure that the edges don't curl. And the same thorough checks of fabric quality and jacket fit that one finds in GFT's large Settimo plant outside Turin can be found, reproduced almost identically, at Riverside.

GFT also changed the ways workers at Riverside are supervised. When GFT took over the company, it kept the top managers—including Riverside's former owner and current COO, Robert Brier—but changed all the frontline supervisors. The company insisted that each supervisor know all the jobs in his or her area, which had rarely been the case in the past, and promoted a number of former operators to supervisory positions.

What is most striking about Riverside is how quickly the new production system was put into place. Even during the first year after GFT purchased the company, when the transformation of the plant was still in process, Riverside produced 10,000 jackets. By the end of the second year, output reached roughly 50,000 and has since grown to the current 70,000 pieces per year. Last year, GFT invested another $400,000 at Riverside to create a new trouser operation that began operating in June. By the end of February 1991, it had produced 32,000 pairs of pants—10,000 more than originally projected. By the end of this year, Riverside will be producing 20% of the clothing that GFT sells in the United States.

What's more, some of the products made at Riverside are among the highest quality garments made in the United States. "As a contractor, our quality was fairly decent," says Robert Brier, "but today we are in a totally different world." Riverside has even begun to make some jackets for an Ungaro collection previously made in Italy. According to Umberto Caccina, the quality of the Ungaro jacket made in Italy and the one made at Riverside is exactly the same. And while production in Italy is more efficient, cheaper labor costs at Riverside and the absence of the duty charged on foreign-made apparel make the final cost of the American-made product 15% to 20% less than that of the jacket made in Italy.

As time goes on, the flows of learning and innovation at Riverside are diversifying. When he first began retooling the Riverside plant in 1987, Caccina used only GFT technology from Turin. Four years later, he still spends 30% of his time in Italy, scouring the GFT system for technology and technical expertise. "Without Caccina," says Brier, "we wouldn't have been able to take advantage of all that GFT has to offer."

Now, however, Caccina is as likely to buy U.S. as Italian technology. He has found that when equipment breaks down, it's better to be an important customer to a local supplier than a small internal unit dependent on a distant bureaucracy in Turin.

And Caccina is always ready to take quick advantage of new opportunities. For example, in 1989 GFT acquired Bäumler, Germany's third largest apparel manufacturer, as part of its insider strategy in the important German market. Caccina is already buying German technology from Bäumler for use in the New Bedford plant. Approval and capital come from New York, not Turin.

Riverside's business relationships reflect a similar complexity. The company is a subsidiary of GFT USA and a supplier to some of its key business units. And yet, through Caccina's rich network in Italy, the company's links to Turin are in some sense stronger than they are to New York. "Riverside is almost like a separate entity that exists away from New York and away from Italy," says Brier. "With Caccina, we've operated as if we were a totally separate company."

After four years in New Bedford, Caccina believes that Riverside has a thing or two to teach to GFT's traditional Italian plants. In his opinion, small, relatively autonomous manufacturing facilities like Riverside, close to the market and able to adapt quickly to change, represent an important model for how GFT needs to organize manufacturing in the future. But sometimes he worries that GFT's manufacturing bureaucracy in Turin doesn't grasp the significance of the plant as an organizational innovation. "Riverside is a successful experiment that, unfortunately, some people in GFT don't take seriously enough," he says. "They don't know the treasure they have here."

Whether GFT as a whole can learn from the Riverside experiment and leverage the local innovation occurring there will depend on the ability of Caccina to transmit what he has learned and the willingness of others to receive it. In this respect, his challenge is similar to that facing Marina Mira d'Ercole, Leopoldo Borzino, or even GFT chairman Marco Rivetti: to be a kind of "organizational designer" who continuously invents and reinvents the company according to the always changing requirements of the business. Indeed, the real location of GFT's designer organization is not so much at the center or even on the periphery but in the minds of its global managers.

3
How Networks Reshape Organizations—for Results

Ram Charan

In a world of increasing global competition and unrelenting change, many companies have been strong on crafting vision and strategy and weak on delivering results. As they struggle to improve their capacity to execute, senior managers use words like trust, teamwork, and boundaryless cooperation to describe the organizations they aspire to build.

Recently a new term—networks—has entered the vocabulary of corporate renewal. Yet there remains much confusion over just what networks are and how they operate. In some companies, networks imply a set of external relationships—a global web of alliances and joint ventures. In others, networks mean informal ties among managers—floating teams that work across functions and maneuver through bureaucracy. Still other companies define networks as new ways for executives to share information, using management information systems, videoconferencing, and other such tools.

I have spent four years observing and participating in the creation of networks in ten companies based in North America and Europe. These companies are clear about why they are creating networks, what networks are, and how they operate. To them, networks are designed to build the central competitive advantage of the 1990s—superior execution in a volatile environment. No traditional corporate structure, regardless of how decluttered or delayered, can muster the speed, flexibility, and focus that success today demands. Networks are faster, smarter, and more flexible than reorganizations or downsizings—dislocating steps that cause confusion, sap emotional energy, and seldom produce sustainable results.

A network reshapes how and by whom essential business decisions get made. It integrates decisions horizontally at the lowest managerial levels and with superior speed. In effect, a network identifies the "small company inside the large company" and empowers it to make the four-dimensional trade-offs—among functions, business units, geography, and global customers—that determine success in the marketplace. It enables the right people in the organization to converge faster and in a more focused way than the competition on operating priorities determined by the imperatives of meeting customer needs and building concrete advantage.

A network is a recognized group of managers assembled by the CEO and the senior executive team. The number of managers involved almost never exceeds 100 and can be fewer than 25—even in global companies with tens of thousands of employees. The members are drawn from across the company's functions, business units, and geography, and from different levels of the hierarchy. Membership criteria are simple but subtle: What select group of managers, by virtue of their business skills and judgment, personal motivations and drive, control of resources, and positions at the juncture of critical information flows are uniquely qualified to shape and deliver on the corporate strategy? Managers who pass these tests become the core network, hold regular meetings, form subnetworks for critical operating tasks, and use conference calls, electronic mail, and computerized information systems to share information.

Networks really begin to matter when they affect patterns of relationships and change behavior—change driven by the frequency, intensity, and honesty of the dialogue among managers on specific priorities. Networks are designed to empower managers to talk openly, candidly, and emotionally without fear, to enrich the quality of their decisions, to test each other's motives and build trust, and to encourage them to evaluate problems from the perspective of what is right for the customer and the company rather than from narrow functional or departmental interests.

Consider three snapshots of emerging networks.

At Conrail, the freight transportation company based in Philadelphia, 19 middle managers drive the company's key operating decisions. The network, called the operating committee, meets for up to two hours on Monday mornings to review and make decisions on a wide range of tactical issues—searching for the right mix of price, delivery schedule, and consistency of service that meets the needs of important customers at low cost and that generates competitive returns in a capital-intensive business. The operating committee is also developing a five-year business plan—a first-of-its-kind analysis meant to generate radical new approaches to important segments of the business. Senior executives join in the dialogues of the operating committee and receive briefings on the evolving substance of the strategic plan, but they neither chair the sessions nor dominate the deliberations.

In Montreal, Royal Bank of Canada, one of the largest and most profitable banks in North America, has embraced management reforms that promise to accelerate and sharpen implementation of its retail strategy. The reforms grew out of a six-month study by a network of 12 managers. They conducted a candid and exhaustive review of the most important operational and human resources issues facing the retail bank. The members, called the community banking team, were not senior executives; rather, they were field officers from across Canada—middle managers with titles such as vice president for retail banking in Alberta and area manager for North Winnipeg. They presented the results of their study, including 17 concrete proposals for change, to a major conference of bank executives and other senior officers in May 1990. Since then, senior management has worked with the group's leadership to implement the plan. One goal is to turn all of the bank's area managers into a network—a close-knit group that routinely shares best practices, learns from each other's problems, and works together to understand the business more deeply.

At the U.K. headquarters of Dun & Bradstreet Europe, a "development network" evaluates and monitors new business-information products and the customization of existing products. For years, D&B Europe has faced the tensions that afflict so many cross-border and cross-cultural organizations. Its computer databases, technical staff, and marketing group are centralized in the United Kingdom—a sensible structure given D&B's position as the only pan-European competitor in its business. But centralization has made it harder and more time-consuming to tailor Europewide products to the needs of local customers and to set priorities among competing projects in different countries.

The development network is designed to make these trade-offs more quickly and more skillfully. It is neither a new layer of bureaucracy nor a means to wrest power from the functional organizations.

Rather, a core of 12 or so key players meet weekly to monitor the performance of the development process, identify barriers, and devise ways to remove them. One of the network's first steps was to create an investment-management function responsible for specifying which projects get done, in what order, and how quickly—in ways that meet the needs of customers and countries but that also reflect corporate goals and strategies. The network reports to the field through a designated representative in each of the 13 country organizations, thus increasing the speed of communication and reducing confusion. Some members act as business project managers, with complete cross-functional responsibility for the success of major projects.

These three companies, along with the others I have observed, share a common understanding of the urgency and volatility of their competitive environment and the need to break with decision making based purely on hierarchical and functional authority. They are also clear about how networks differ from teams, cross-functional task forces, or other ad hoc innovations designed to break hierarchy. First, networks are not temporary. Most task forces assemble to solve a specific problem and then disband and return to business as usual. They do not sustain change in the behavior of the organization. Members of a network, on the other hand, identify with it and with each other. The frequency and honesty of their dialogues reshape personal relationships. Continuous practice over a sustained period of time builds a shared understanding of the business. Networks even affect how their members move through the company. Managers' performance and promotability is evaluated with respect to their contribution to the network and sometimes by the network itself.

Second, unlike most teams and task forces, networks do not merely solve problems that have been defined for them. Networks are dynamic; they take initiative. They become the vehicle to redirect the flows of information and decisions, the uses of power, and the sources of feedback within the hierarchy. They become a new way of doing business and a new operating mechanism for individual managers to make their presence felt.

Finally, networks make demands on senior management that teams and task forces do not. CEOs and their direct reports no longer define their jobs as making all substantive operating decisions on their own. Rather, their primary job is shaping the processes and personal relationships that allow other managers, the members of the network, to make decisions. To be sure, top managers still set goals and make all kinds of decisions—about personnel, resource levels, acquisitions and divestitures, upgrading technical competences. But they must also become more adept at diagnosing the behavior of their organization, building relationships among key managers, modifying measures and rewards, and linking all of these "soft" changes with the company's economic performance.

This article examines the process of building and sustaining networks in large organizations—a process that begins at the top. Senior managers work as change agents to create a new "social architecture" that becomes the basis of the network. Once the network is in place, they play at least three additional roles. First, they define with clarity and specificity the business outputs they expect of the network and the time frame in which they expect the network to deliver. Second, they guarantee the visibility and free flow of information to all members of the network and promote simultaneous communication (dialogue) among them. Finally, they develop new criteria and processes for performance evaluation and promotion that emphasize horizontal collaboration through networks. They openly share these performance measurements with all members of the network and adjust them in response to changing circumstances. Let us explore these areas one at a time.

Social Architecture and Change

The foundation of a network is its social architecture, which differs in important ways from structure. Organizational structure refers to the systems of vertical power and functional authority through which the routine work of the organization gets done. Social architecture refers to the operating mechanisms through which key managers make trade-offs and to the flow of information, power, and trust among these managers that shapes how those trade-offs get made. Social architecture does not concern itself merely with who is in the loop or the process by which the loop forms. Social architecture concerns *what happens* when the network comes together—the intensity, substance, output, and quality of interactions—as well as the frequency and character of dialogue among members on a day-to-day basis.

A robust social architecture does not imply absolute harmony among peers. Indeed, the single most important role of networks is to surface and resolve conflict—to identify legitimate disagreements between functions, regions, and business units and to make difficult trade-offs quickly and skillfully. A robust social architecture encourages members of the network to become mature and constructive in their approach to conflict, to direct their energies toward the substance of disagreements rather than toward personal clashes and politics, to search for creative solutions rather than to look over each other's shoulders, and to identify new challenges.

Senior management drives the process of building a new social architecture. The first step is design. The CEO must identify the important decision makers in the organization, assemble them into a network, and communicate it throughout the company. This can be a sensitive step since the process of including some managers in a network necessarily means excluding others. And membership seldom relates directly to hierarchy or seniority. Indeed, the criteria are specific to each company, a function of the unique challenges facing the business and the strengths and weaknesses of the managers themselves. The basic objective, though, is to find the right mix of managers whose business skills, personal motivations, and functional expertise allow them to drive the enterprise.

At Conrail, for example, the core network comprises fewer than 50 managers drawn from three layers below the CEO. The operating committee is a subset of this core network. Explains James Hagen, Conrail's CEO: "There are no more than 25 people in this company whose close, horizontal collaboration will have a dramatic impact on the bottom line. There are the seven assistant vice presidents in the marketing department responsible for our lines of business—steel, autos, intermodal, and so on. There are the six general managers responsible for railroad operations in different parts of our service territory. There are some key people at headquarters—the chief mechanical officer and the chief engineer, the head of customer service— as well as the senior management group. On their own, none of these managers can move the business decisively. As a network, they are already making a visible difference."

As the social architecture begins to change, the performance of managers whose behavior hurts the network becomes visible. This raises a second crucial role for senior management—dealing with mismatches by reassigning the problem executives. Networks play to the best instincts of people. Most middle managers don't resist change. They want to cooperate, share information, be open and secure in their interactions with peers. Networks quickly surface people of exceptional competence, informal leaders whose talents have been hidden behind functional or hierarchical walls. They also reveal managers whose business skills or personal style are mismatched with the needs of the network, people who simply can't make the change to a new way of doing business. These managers can have a toxic effect on the rest of the group. At nearly every company I have studied, at least one senior executive (but never more than two) has had to be reassigned or dismissed because of an inhibiting presence and unwillingness to change. With respect to social architecture, one bad apple *can* spoil the whole bunch, especially if that person is in a position of vertical power in the organization.

Third, and most important, building the social architecture requires an intense and sustained focus on the fundamentals of the business rather than abstract appeals to culture, teamwork, or values. Companies do not build networks so that managers will "like" one another or behave like "family." Networks are designed to develop professional trust and empathy and a richer and more widely shared understanding of the specifics of the business. No generic change program imported from the outside can generate such understanding. When network members identify real business problems, diagnose them together, create a broad and common base of specific information, and reach conclusions that reflect the pressures and capabilities of multiple functions and geographic units, they become more skillful at making trade-offs, and more trusting of one another. The keys are immersion, concentration, practice, and the simultaneity of information flows.

The rise of networking at Conrail illustrates the change process associated with building a new social architecture. The operating committee, which officially took shape in late 1990, is the railroad's core network for profitability—the central operating mechanism through which important business trade-offs get made. All told, though, 46 managers from around the company worked for two years to build the trust and confidence necessary to support a departure as radical as the operating committee. The goal was to create a new social architecture among these managers—a real challenge given Conrail's past.

Conrail is an unlikely candidate for management innovation. No industry compares with railroads for its legacy of rigid hierarchy and authoritarian management. And no major railroad compares with Conrail for its commitment to relentless and painful cost cutting and retrenchment. Since 1976, when Conrail was created from the remains of the bankrupt Penn Central and several smaller lines, employment has fallen from roughly 100,000 workers to fewer than 28,000.

These are not ideal circumstances in which to build trust and professional empathy. James Hagen, who became CEO in May 1989, understood that he could not simply assemble a group of middle-level managers into a network and expect them to erase decades of learned behavior. The change process began with the selection of two core groups of managers. Conrail has 450 or so top-line managers, people whose decision-making authority directly or indirectly affects operating income. Top management selected fewer than 50 from this group to serve in the two networks. The senior planning team consisted of 13 executives, essentially the company's top officers, although it did deliberately include two lower level managers. The second network, the strategy managers group (SMG), consisted of about 35 executives from many functions and departments. This is where judgment became important. Top management developed a statement of its selection criteria for the SMG: "The smallest working group whose interlinking can significantly affect both the operation and selling of our basic services." In practice, the criteria meant that SMG membership was weighted toward field managers rather than corporate staff and toward field managers with direct responsibility for pricing service, meeting customer demands, or running the railroad. (Later, as the new social architecture began to evolve, some staff members shifted to line jobs.)

The creation of the network was followed by a series of initiatives designed to change the character of the dialogue and interactions among the members. Some of the techniques were simple, almost quaint. For example, the company published a directory of the SMG members complete with photographs and descriptions of career histories and personal interests. The directory reinforced the SMG as a living network rather than a one-time task force or team. It helped create an identity for a group that had never assembled together before. In fact, several members, despite their complementary roles in the company, had never even met one another.

More substantively, the two groups began a series of meetings, first separately, then together, to diagnose the future of Conrail. At all times, both groups focused tightly on the business itself—there was no wilderness experience, team building, or other generic exercises. For example, the SMG identified nine priority issues and created small teams (or subnetworks) to study them and report back with action-oriented recommendations. The projects included some of the most politically sensitive problems inside the railroad: work force reduction, billing quality, and managerial rewards and evaluation.

These teams became the building blocks of the new social architecture. Small groups of managers, most of whom had barely spoken to each other before the creation of the SMG, devoted hundreds of hours to analyzing real problems and convincing the senior planning team to endorse their recommendations. This investment of time (over and above each member's day-to-day responsibilities) and the total immersion that the subnetworks required created personal bonds of enormous strength.

What's more, the subnetworks embraced rather than avoided conflict—a critical test of the members' good faith and capacity to air differences openly. For example, one group studied customer service at Conrail, specifically, whether and how to consolidate the three separate departments and ten different locations responsible for resolving billing disputes, tracking freight, and otherwise interacting with customers. Customer service had been a sensitive issue for many years; past efforts to consolidate it had been squelched after painful and demoralizing turf battles.

The group faced those turf battles squarely. Each of the six members had a direct or indirect functional stake in customer service: the general manager for information systems, the assistant vice president for customer service, a regional general manager, the assistant vice president for labor relations, the corporate treasurer, and the general manager for stations. Needless to say, a recommendation to consolidate would require compromise among these previously warring factions. But there was more. A move to consolidate would also require two members of the group to eliminate their own jobs; their positions would become redundant. The fact that these two managers took that step and spent several months with unclear prospects about their futures was a powerful statement of the group's capacity to resolve conflict.

The subnetwork also reckoned with conflict vis-à-vis senior management. During the course of their analysis, the group learned that certain senior executives were working behind the scenes to oppose consolidation. At an interim presentation to the SMG, the team members decided to take a stand. They proposed that they dissolve rather than make recommendations destined to fail. The rest of the SMG insisted that the subnetwork stay together and make its presentation to the senior planning team. It did, and after much debate, top management agreed to make the change.

In November 1990, Conrail announced it would consolidate all customer service activities at a new facility near Pittsburgh. To the outside world, this was a reasonable and welcome step to improve service. To the SMG, it was a major victory and a sharp break with the past—the sort of development that builds professional trust, unleashes energy, and, over time, changes behavior.

The Power of Specificity

As the Conrail experience demonstrates, networking represents a decisive break with the past. Although it plays to people's best instincts, it is a demanding journey that requires managers to unlearn attitudes and behaviors reinforced over decades. No amount of general debate over business strategy and vision can build the trust and confidence required for new behaviors to emerge. All too often, managers invest huge amounts of time and energy reaching agreement on vision without investing in the truly hard work—becoming aligned on the nitty-gritty trade-offs and time pressures required to deliver on the vision. Agreement without alignment seldom changes behavior. Instead, it generates frustration, cynicism, and complacency.

That's why senior executives must define with clarity and specificity the business outputs they expect of the network and the time frame (usually less than six months) in which they expect the network to deliver. Such precision forces managers to reckon with the day-to-day realities of the business. Meanwhile, the dialogue and debate required to shape the targets builds commitment and confidence. Specificity allows the members of the network to see that they are making real change and that the change is linked directly to the improved economic performance of the organization. The more visible and persuasive this evidence, the more intense the personal commitments of the members to expanding the initiatives of the network. Over time, making visible progress on economic-priority items generates emotional energy and builds commitment.

The remarkable effectiveness of the Royal Bank of Canada's community banking team is a case in point. In less than one year, a network of 12 managers, most of them middle managers drawn from the field, made a mark on some of the most sensitive management issues in the retail operation: performance measures, spans of control, training and development. How did it succeed? By producing concrete deliverables, developing techniques to measure the organization's support for their proposals, and setting specific dates for implementation. (See "The Power of Specificity—Royal Bank of Canada.")

The Power of Specificity—Royal Bank of Canada

In the spring of 1990, 350 Royal Bank managers gathered in Montreal for a first-of-its-kind (and size) leadership conference. There the community banking team, a 12-member group assembled by chairman Allan Taylor the previous November, unveiled its six-month study. The team's charter was ambitious: to evaluate how well the bank was executing a bold new retail strategy and to recommend without limitations new initiatives to drive execution.

The strategy had been in place for several years. Its main organizational thrust was to increase the authority and prominence of the bank's area managers, senior field officers responsible for the performance of groups of bank branches. Top management was genuinely committed to transforming how operating decisions were made. Indeed, the plan could have been a primer on translating the abstract rhetoric of "empowerment" into tangible management principles. It emphasized the importance of personalized leadership by area managers, local market intelligence, interactive planning driven by the field, and local autonomy to provide area managers with the flexibility they needed.

Precisely because the retail strategy was such a departure and the team's charter so ambitious, there was a risk that the final report would disappoint, especially since no business crisis was evident. Royal Bank is the largest bank in Canada and the third largest in North America. It is also one of the most profitable. Its 1990 net income was nearly $1 billion, more than any U.S. commercial bank, and its return on assets and equity consistently rank it high among North America's leaders. Retail is at the heart of this outstanding performance; Royal Bank has more than 7.5 million customers, 1,600 branches, and 3,100 cash machines.

But the report does not disappoint. It is the most candid analysis of Royal Bank's retail operations ever developed. Within a year of its completion, the bank had implemented nearly all of the team's proposals. The work of 12 managers, most of them drawn from the field, it is reshaping some of the most sensitive management issues in the retail operation: performance measures, spans of control, training and development. Why? One reason stands out: by producing specific deliverables, setting precise dates for implementation, and developing ways to measure progress, the community banking team embraced the power of specificity and used it to advantage.

The team's written report ran less than 100 pages, and its presentation to the spring conference (along with group processing of the presentation) lasted three hours. Packed into this limited space and time was a penetrating analysis filled with evaluations of particular incidents, documents, and success stories. There were no abstract calls for greater empowerment, more resources, or a deeper commitment to customer service.

Each of the team's proposals (there were 17 in all) included numbers, targets, or specific language that left no room for ambiguity or misinterpretation. For example:

Limit the number of branches under the control of an area manager to between 7 and 12.

Create formal mechanisms (conferences, newsletters, computer systems) to allow area managers to share expertise and competitor intelligence more systematically.

Guarantee 15 days of training per year for existing area managers and launch Area Manager University, an intensive series of courses and training modules. Provide resources so that all new area managers would be graduates of the university by the spring of 1991.

Institute standard, minimum autonomies for all area managers, including the right to hire, transfer, and evaluate employees below a certain pay grade, and the right to freely substitute between certain expense categories within the overall budget constraints. Negotiate even greater autonomies with high-performing area managers.

Of course, the very specificity of these recommendations meant they were likely to be controversial. So the community banking team, working with the top managers who designed the leadership conference, created special group-processing techniques for immediate feedback from the group of 350. For example, the group was divided into sets of ten, each set reflecting a mix of hierarchy, function, and geography. These sets debated the report and were invited to offer feedback.

Here too clarity played a role. The team did not merely ask for general reactions. Rather, personal computers allowed the 350 attendees to "vote" on a series of questions designed to measure their enthusiasm about, confidence in, and commitment to the proposals. The voting was confidential, but the results could be tabulated by function, geography, or hierarchical position. The community banking team (and the entire leadership conference) could see immediately whether managers in Western Canada reacted differently from those in Montreal or whether headquarters executives were more strongly in favor of a proposal than field managers.

This detailed feedback was a vital source of information for the community banking team. It also energized and built the commitment of the audience. Never before had such a large group of executives been privy to such candid, detailed analyses and proposals.

The final step was implementation. The day after the leadership conference, the community banking team held an all-day session with Royal Bank's top 50 executives. This group of 50 had been meeting on a regular basis for several years, working to build their own horizontal collaboration and trust. By securing the collective commitment of these executives, the community banking team acquired powerful allies. Moreover, for each of the 17 proposals, at least two managers—one from the community banking group and one from the group of 50—accepted deadlines and personal responsibilities for implementation. Successful implementation also became part of the annual objectives by which the senior executives were evaluated.

The results have been encouraging. By the middle of 1991, all geographic areas had been broken down into units small enough that an area manager can visit every branch at least once a week. Area Manager University is up and running, and the first graduates have completed their course work. The retail bank has implemented standard "key result areas" so that all area managers know the criteria on which they will be measured. Nearly two-thirds of the area managers have negotiated extended autonomies with senior management, and negotiations continue with the remaining area managers.

The work of the community banking team is just a beginning. The ultimate goal is for area managers across Canada to function as a horizontal network: sharing expertise, transferring best practices, learning from mistakes. In fact, all 175 area managers will come together in November 1991 for their first-ever meeting. The concrete, measurable, and substantive achievements of the community banking team will serve as an encouraging and energizing backdrop.

The Conrail operating committee is another example. After only six months, the network can point to concrete achievements. For example, the introduction of a computer-based simulation model is helping to redesign the company's huge intermodal (truck-to-train) business. But all 19 members of the network are clear about their ultimate goal—to move the railroad's operating ratio to 80% from the 1990 level of more than 87%. This is not an arbitrary target. The ratio, which measures operating expenses as a percentage of revenues, is the company's central indicator of profitability. Conrail needs a ratio of 80% for earnings to exceed its cost of capital and thus for the railroad to remain self-sustaining as a financial enterprise.

Every member of the operating committee (and the broader group of 46) understands the logic behind this target and the urgency of hitting it. The group monitors the ratio as a way to track its progress, and senior management uses specially designed feedback instruments to monitor the network's confidence that it will achieve the target. The fact that the ratio has begun to move even under adverse economic circumstances (most recently the ratio was at 84%) has been an enormous source of emotional energy and confidence—evidence that the network has begun to shape the company's economic performance.

Richard Archer, president of Dun & Bradstreet Europe, has chosen to personalize the D&B network's commitment to clear performance targets. The company's development subnetwork is one of several that Archer has assembled to meet the new cross-border competition in Europe. The challenges facing D&B are similar to those facing all companies struggling to globalize their operations: How does the organization strike the right balance between the power of centralization and the responsiveness of decentralization? How do managers balance the demands of cross-border customers, who expect uniform service at the same price everywhere, with the demands of national customers in what remain very different markets? What makes D&B Europe noteworthy is the clarity with which Archer and the network are addressing these issues.

The network-building process began in January 1989 and, at the time, involved 45 continental managers. Through a series of meetings, this group refined the company's strategic direction and business priorities. Each session strove to develop greater clarity than the one before it, and by late that year, Archer was able to develop a mission statement for D&B Europe. In early 1990, at a three-day workshop, he shared the statement with 98 top managers across all countries and functions, and revised the statement based on their feedback. At the end of that session, he assembled three cross-border project teams (not unlike Conrail's SMG subnetworks) to attack the highest priority issues facing the organization. Finally, drawing on the project reports and the revised mission statement, Archer worked with ten senior managers to craft a document that communicates the tangible and measurable results the network is expected to produce. He insisted that the final document run no more than one page—a discipline that forced even greater clarity and precision—and then published it as his *personal* goals for 1991.

Again, the development subnetwork is a case in point. The network is clear about its goals: reduce lead times by 50%; eliminate 80% of the old system's "process loss"; respond to all development requests (usually modifying existing products) within ten days. The results have been equally clear. Overall lead times for new products have been reduced from one year to six months. The D&B technical center has cleared a backlog of 100 requests from the country organizations; a country manager who puts in a development request can now expect a response in less than three weeks. In 1991, for the first time, D&B Europe agreed to a detailed schedule for the implementation of new products. To date, all important deadlines have been met.

To be sure, the process of settling on meaningful specificity in time-based goals and outcomes is tedious and emotionally demanding. Many managers, especially senior managers, find it intellectually un-appealing—less exhilarating, say, than crafting a stirring vision statement or a bold corporate strategy. (As one CEO noted, "There is a fine line between vision and hallucination.") Specificity requires dialogue, repetition, negotiation, candor, practice. Over time, though, these are the investments that generate alignment on and commitment to the details of execution. And it is precisely the speed and quality of exe-cution—the trade-offs among functions, business units, geography, and customers—that the network is designed to change.

Information—Visible and Simultaneous

No organization of two or more people can function without infor-mation. Indeed, in any organization, the character of information flows is one of the most critical variables determining the speed and accuracy with which decisions get made—and thus the quality of execution. Two companies with identical structures will behave very differently based on the answer to a simple question: Who receives what information when, and why do these people (as opposed to others) receive it? In most large companies, the flow of information remains incomplete and sequential and thus prone to distortion and manipulation.

In a network, especially a global network that extends across bor-ders, information must be visible and simultaneous. When members of the network receive the *same* information at the *same* time, and receive it quickly, business decisions take on a different character. Disagreement over substance no longer generates damaging personal conflicts or organizational politics; personal conflicts arise when information flows are selective or secret. More to the point, disagreements themselves are reduced. Most members of a network, when faced with an impartial set of facts, tend to arrive at roughly the same business options for a given set of goals. The visibility and simultaneity of information improve the quality of decisions.

Some of what I mean by information is hard data on performance, the operating and financial indicators that capture the pulse of the company. At Conrail, for example, all 46 members of the network are connected through a management information system called Commander. They receive daily reports on critical business indicators: traffic levels, carloads, revenues per ton, and many others. Sharing this data openly and immediately and using electronic messages to allow people to evaluate the information enrich the business perspectives of network members, keep them updated and focused on the company's performance, and allow managers from different parts of the company to offer their unique perspectives on specific business developments.

But data are in many respects the least important dimension of information. The network must also share openly and simultaneously each member's experiences, successes, and problems, soft information that can't be captured in databases and spreadsheets and that remains hidden for as long as possible in most traditional organizations. This is the kind of sharing that builds trust, empathy, and secure relationships. It also broadens the participants. They begin to see the organization through multiple viewpoints and understand more instinctively the pressures, challenges, and capabilities of functions and business units outside their own traditional boundaries.

Consider the case of Armstrong World Industries. In late 1989, Armstrong created five cross-border, cross-functional networks to begin the process of globalizing its businesses. Armstrong is one of the few companies with a strong international presence in many of its product lines, but it can hardly be considered a model global organization. It is a product of its conservative roots in the Pennsylvania Dutch country. Most decision-making power resides in its Lancaster headquarters, and virtually all R&D is still conducted there.

These "global teams"—one each for flooring, building products, insulation, gaskets, and textiles—have been working for more than a year to increase cross-border collaboration and to make more effective decisions on global-local trade-offs. They have held meetings outside the United States, identified business priorities, and begun to solve real problems.

In the last 18 months, the building-products network has visited all of the group's factories in Europe, developed sourcing plans to redesign the flow of products between Europe and the United States, and encouraged Lancaster R&D to tailor new products for the Asian market. (The new R&D responsiveness has increased Armstrong's share of the Korean building-products market from virtually zero to 15%.) The network has also encouraged lower ranking managers to emulate network interactions. Plant managers in the building-products group have already held two global conferences—intensive sessions in which they share technical information and operational insights.

Still, network members understand that much work remains to be done and that their work requires frequent contact and dialogue. Yet these managers are based thousands of miles from each other, travel widely, and find it almost impossible to gather physically more than three or four times each year. So they have devised a simple technique to share and react to hard and soft information—regular conference calls in which all members of the network participate all the time. This may seem a modest technique, and, technically speaking, it is. But it works because the members have learned to trust each other—and because sharing and reacting to new information fully and frequently enhances that trust.

For example, the ten-member building-products network includes managers from Pennsylvania, Europe, and the Pacific. At least every other Monday at 7:00 A.M. Lancaster time (which turns out to be a convenient time for all ten managers worldwide), the network participates for at least an hour in a global conference call. Henry Bradshaw, an Armstrong group vice president and the leader of the building-products network, describes the substance of the calls: "We talk about business conditions. We talk about competitors. We talk about service, which is especially important in the Pacific. We talk about particular orders. We talk about new products. Europe wants to know where the labs are with a new kind of ceiling board that they think will be a big hit in their market. Simple as they sound, these conference calls have been very effective. Before we created the global network, I didn't know most of the guys on it. And we had complicated communication channels; important information got lost. Today we're on a first-name basis. The more we talk, the more we want to talk. The more information we share, the more natural it becomes to share it."

Armstrong also shares openly and simultaneously a third critical kind of information—how the network evaluates its own performance. Recently, senior management polled each member of the network with a specially crafted survey designed to measure how behavior is changing and whether the group is producing tangible results. The surveys had four core themes: progress on global business projects, the quality and frequency of global communication, changes in individual mind-sets, and the global team process. The questions were direct: How often do you communicate with your team members? To what extent has your ability to see a broader global picture of the business grown? To what extent has your behavior been inhibited because of the reward system? What is the nature of conflict resolution during team meetings? What steps have you taken personally to build the social architecture? The answers almost always involve numerical rankings or choices among specific alternatives—a way to keep the information concrete and comparable over time as management repeats the exercise.

This kind of detailed evaluation often allows top management to act more quickly or more decisively than it otherwise might. For example, Armstrong recently announced a corporate streamlining that simplifies global reporting relationships. This streamlining was a direct outgrowth of the networks' self-evaluations. The more the global networks worked on concrete problems, the more convinced they became that the hierarchy needed simplification. The feedback surveys documented wide agreement and a need for urgency, so top management took action. In effect, the networks helped reshape Armstrong's corporate structure.

Armstrong executive vice president Allen Deaver emphasizes this spillover effect: "The creation of the global teams and the measuring and sharing of members' perspectives on their work and the company helped us make better decisions about people and organization. This change is designed to facilitate the work of the teams. It is natural, logical, and evolutionary—and perceived as such. So we expect it to generate energy rather than confusion. It is perfectly in line with policies the teams themselves endorsed."

Sharing information openly, visibly, and simultaneously is one of the most important dimensions of sustaining a network. Over time, the free flow of information allows networks to become self-correcting. New information inspires debate, triggers action, generates offers of assistance from network members—without instructions from above. Networks cannot altogether abolish personal and departmental rivalries. But sharing information openly makes them visible and encourages people to redirect their energies in constructive directions. It eliminates the need for checks and checkers and becomes a central building block in improving managerial productivity.

Measuring Performance: New Criteria, New Systems

The single most important lever for reinforcing behavior in networks is evaluation. Every manager, regardless of position or seniority, responds to the criteria by which he or she is evaluated, who conducts the review, and how it is conducted. In most organizations, even those committed to cross-functional collaboration, reviews are still based on performance against departmental budgets or functional goals and are still conducted vertically.

These traditional reviews are sharply at odds with networks. Vertical reviews encourage turf mentality. Functional reviews promote narrow vision of the business and discourage horizontal collaboration. For a network to thrive, top management must focus on behavior and horizontal leadership: Does a manager share information willingly and openly? Does he or she ask for and offer help? Is he or she emotionally committed to the business? Does the manager exercise informal leadership to energize the work of subnetworks?

Admittedly, these are difficult criteria to quantify, which is why they seldom figure into performance reviews. But in a network, they become the central measures of performance, the vital criteria for assigning increasing general management responsibility. Senior management must find ways to make these criteria precise, to demonstrate that they are the basis for rewards and promotion within the organization, and to share them among all members of the network. In so doing, management reduces the old-boy-network style of vertical promotion politics.

Royal Bank of Canada has taken the lead in devising new performance metrics that reflect the behavioral imperatives of networks. Top management has developed a one-page statement that specifies the five new criteria by which the bank's top 300 executives will be evaluated. These criteria are crucial ingredients in succession planning and career development. And they are noteworthy precisely because they emphasize behavior and mind-set rather than functional expertise. Although the document itself must remain confidential, these excerpts suggest the qualities it emphasizes:

A strong business-profit orientation: instinctively thinks customer needs, customer service; understands the anatomy of the economic structure of the business; strong innate instincts for making money.

Demonstrated ability to accept accountability, assume leadership and initiative: raises standards constantly as the environment changes; by personality and chemistry, is open and secure in assuming the initiative in building leadership, without horizontal power or authority; believes in sharing information and engendering trust; less control mentality, more empowering mentality; team builder.

Demonstrated record for making a "qualitative shift" and impact on the bank: has shown the vision and courage to change things, not just run things; willingness to experiment.

Astute in the selection of people: demonstrated evidence that this person has the judgment and the security to select and build a team of superior people; willing to cut losses.

Intellectual curiosity and global mind-set: has the mental makeup for learning continuously about global developments, technology, etc. from the outside world.

Royal Bank has also devised an ambitious evaluation program called the "leadership review process" to put these criteria into practice. The new process is a sharp break with the past. Under the old system, a committee of senior executives (the chairman, the president, four other top executives, and the director of personnel) convened twice a year to review the performance of the bank's top 200 or so managers. Each was assigned to one of three lists—the "A" list, the "B" list, and the "C" list—that reflected their perceived career potential at that point in time. "A" list members were high performers capable of moving into top management; "B" list people were promotable but not outstanding; and "C" list members were not promotable.

This process had the virtues of simplicity and efficiency, and top management was largely satisfied with the outcomes. But there were at least two serious weaknesses, especially for an organization committed to collaboration through networks. First, knowledge of a manager's leadership and personal style (as opposed to quantifiable business results) was limited to the handful of top executives in the room, few of whom had worked directly with the managers under review. Also, the standards of evaluation, especially with respect to behavior, were more intuitive than explicit. The review team could see in black and white whether a manager's department was running ahead of or behind budget and by what percentage. But how could it evaluate with any clarity how managers shared information, how eagerly they worked to solve problems outside their domains, how effectively they led without authority?

The new system uses a four-step process to overcome both weaknesses. Here is an abbreviated description. First, the manager under review writes a statement of career aspirations and a self-appraisal against the performance criteria. Next, the manager submits a list of colleagues to be interviewed by a trusted executive about his or her performance. (Top management's selection of the interviewer is uniquely important; he or she must be widely regarded as honest, apolitical, and seasoned in business judgment.) The list must include at least 7 names (3 peers, 3 subordinates, and 1 boss) but can go as high as 15. The interviews run for an average of one hour and, like the criteria themselves, are meant to be subjective and to focus on behavior. These are not mere note-taking sessions by a human-resources staffer or scripted conversations to rank-order executives along a few obvious dimensions. The interviewer is a savvy, senior-level player with keen insights into other executives. He or she probes to elicit each individual's opinions and thoughts, always in the context of the business.

Next, the interviewer prepares a one-page profile of the executive's leadership capabilities. This profile, along with the self-assessment, goes for final review to the president and the chairman. In "feedback sessions," the interviewer discusses the review with the executive and allows him or her to express reservations and reactions.

The leadership review process represents an enormous commitment of time and resources. Indeed, after two years, the bank has still not completed its evaluations of all 200 top executives. But think of its powerful impact on behavior and motivation—even beyond the 200 executives. The dialogue through which top managers devised the performance criteria was itself a valuable learning exercise for them and for everyone who sees the criteria. For the first time, standards of effective behavior have been made visible and concrete. Moreover, the new evaluation system influences the values and behavior of managers interviewed about the performance of their peers and superiors. These managers (many of whom get interviewed about more than one executive) listen to the questions, formulate judgments about individual leadership, decisiveness, and accountability, and become more aware of their own values and behavior in the process. Finally, the promotion of executives based in part on the review—people who might not have been promoted under the old system—sends a visible and powerful message throughout the bank about how management values horizontal leadership and behaviors that sustain the network.

Networks offer a promising alternative to the two-dimensional lens of strategy and structure through which most senior managers evaluate their organizations. By addressing the character of the interactions among a select group of managers, rather than the formal relationships among hundreds or thousands of managers, networks begin to give substance to the values that CEOs so often invoke—trust, teamwork, empowerment—but so seldom deliver. Over time, the members of the network influence values and behavior both above and below them in the larger organization. In this way, networks become an essential building block in the creation of boundaryless organizations.

The companies I have observed understand that their networks remain in the early stages of development. But their success thus far—at Conrail, Dun & Bradstreet Europe, Royal Bank of Canada, Armstrong World Industries, and the other companies I researched— offers evidence of the power of this approach and its impact on real economic value. The performance of networks should also go a long way to overturning the fashionable (and all-too-convenient) perception of middle managers as obstacles to corporate change. Middle managers are not born defensive and narrow-minded. They learn their behaviors from those above and around them. Most people want to cooperate and collaborate. They would prefer to innovate than to block. By forging a strong set of relationships and values, networks reinforce managers' best instincts—and unleash emotional energy and the joy of work.

Appendix: A Note on Research

For the last four years, I have observed and participated in the creation of networks in ten companies based in North America and Europe: Armstrong World Industries, Blue Cross/Blue Shield of Georgia, CIGNA, Conrail, Consolidated Natural Gas, Dun & Bradstreet, Du Pont, General Electric (certain business units), MasterCard International, and Royal Bank of Canada.

Unlike most academic research, my research has been conducted *as events took place*. I worked with management to improve group-processing techniques, to design feedback surveys to measure progress, and to encourage openness and candor in all discussions. By participating in critical meetings and remaining in close touch with the participants, I was able to observe developments "on-line" and test reactions with all participants. This on-line dimension is important. After-the-fact surveys and interviews, the traditional academic research tools, seldom provide the range of perspectives required to understand individual and group behavior in large organizations. As with networks themselves, research benefits from the power of specificity.

4

The CEO as Organizational Architect: An Interview with Xerox's Paul Allaire

Robert Howard

As chairman and CEO of the Xerox Corporation, Paul Allaire leads a company that is a microcosm of the changes transforming American business.

With the introduction of the first plain-paper copier in 1959, Xerox invented a new industry and launched itself on a decade of spectacular growth. But easy growth led Xerox to neglect the fundamentals of its core business, leaving the company vulnerable to low-cost Japanese competition. By the early 1980s, Xerox's market share in copiers had been cut in half, and return on assets had shrunk to 8%.

In the mid-1980s, Xerox embarked on a long-term effort to regain its dominant position in world copier markets and to create a new platform for future growth. The first step: use the techniques of total quality to re-engineer the fundamentals of the business—product development, manufacturing, and customer service. Thanks to the company's Leadership through Quality program, Xerox became the first major U.S. company to win back market share from the Japanese. And return on assets has increased to 14%.

Since becoming CEO in 1990, Allaire has moved quickly to take Xerox's corporate transformation to a new level. He has redirected the company's strategy to position Xerox as "the document company," at the intersection of the two worlds of paper-based and electronic information. And Allaire has guided the company through a fundamental redesign of what he calls the "organizational architecture" of Xerox's document-processing business, which accounts for roughly 77.5% of the company's $17.8 billion in 1991 revenues.

Many companies are reorganizing to cope with new competitive realities, but few CEOs have approached the process of organizational redesign as systematically and methodically as Allaire has. Over the past two years, he has created a new corporate structure that balances independent business divisions with integrated research and technology and customer operations organizations. He has redefined managerial roles and responsibilities, changed the way managers are selected and compensated, and renewed the company's senior management ranks. And he has articulated the new values and behaviors that Xerox managers will need to thrive in a more competitive and fast-changing business environment.

Allaire comes to this challenge with a deep knowledge of the Xerox organization. Since joining the company as a financial analyst in 1966, he has served in a variety of senior management positions at both Rank Xerox, the company's European joint venture, and at Xerox headquarters in Stamford, Connecticut. In addition to his position as CEO, he was elected chairman of the board in May 1991.

The interview was conducted at Xerox's Stamford headquarters by HBR senior editor Robert Howard.

HBR: In the past decade, companies across the economy have been decentralizing, downsizing, and flattening. What makes Xerox's reorganization different from what other companies are doing—or, for that matter, different from what Xerox has done in the past?

Paul Allaire: When most companies reorganize, usually they focus on the formal structure of the organization—the boxes on the organizational chart.

Typically, top management just moves people around or tries to shake up the company by breaking up entrenched power bases. Rarely do senior executives contemplate changing the basic processes and behaviors by which a company operates.

Until recently, Xerox was no different. In the 1980s, we went through a number of reorganizations. But none of them got at the fundamental question of how we run the company.

The change we are making now is more profound than anything we've done before. We have embarked on a process to change completely the way we manage the company. Changing the structure of the organization is only a part of that. We are also changing the processes by which we manage, the reward systems and other mechanisms that shape those processes, and the kind of people we place in key managerial positions. Finally, we are trying to change our informal culture—the way we do things, the behaviors that drive the business.

In fact, the term "reorganization" doesn't really capture what we are trying to do at Xerox. We are redesigning the "organizational architecture" of the entire company.

What is happening in your business that makes this more fundamental change so necessary?

Our business is rapidly becoming more competitive—and more complicated. Xerox grew from a high-tech startup to one of the country's biggest corporations in large part because we were the only game in town. Then in the 1980s, we faced our first serious competition from the Japanese. This posed a real threat to the company, but thanks to our quality process, we have responded to that threat extremely well.

However, the competitive challenges don't stop there. Xerox is now in the midst of a technological transformation that is revolutionizing our business. It is changing the skills our employees need, the competitors we face—and, indeed, the very nature of the business we are in.

How so?

Our traditional business is light-lens copiers and duplicators, fairly sophisticated electro-optical mechanical devices. Increasingly, they incorporate many computer systems—to control for copy quality, for instance—but the guts of the machine are electro-optical and mechanical. These are stand-alone devices.

With the evolution of digital technology, however, and its rapid reduction in cost, the light-lens element in a copier can now be replaced with a scanning device that digitizes the information on the page. Once you've captured that image electronically, you can do all kinds of things with it in addition to just making a copy.

Take the example of our new digital color copier. Once you scan a document into the copier, you can change the colors on it. Perhaps it's an old or damaged document. Well, you can enhance it and clean it up. You can also edit the document and even add photographs. You can merge electronic information—coming from a personal computer or a mainframe or over the network—with paper information. And once it's complete, you can send your new electronic document to be printed somewhere else. You can store it electronically as well, so you no longer have to deal with cumbersome paper files. Our traditional "stand-alone" copying devices are fast becoming components of complex, digital document systems.

What does this technological transformation mean for the business?

Enormous opportunity—and equally enormous challenges. In this new technological environment, it's not enough anymore just to make and sell discrete products. Rather, we need to offer our customers distinctive capabilities that will have a major impact on the way they do business.

As the worlds of paper-based information and digital information merge, we have redefined our strategic direction to focus on "the document." We believe that documents—whether in electronic or paper form—are key to making organizations more efficient. By focusing on managing documents and using them more effectively, Xerox can help our customers improve productivity.

What are the implications of that strategy for Xerox as an organization?

We have to change. For one thing, our people need new skills. Product designers now have to be far more knowledgeable about computer systems and electronics. They have to know how to function in the new "systems" environment.

But probably the greatest impact is on our relationship with customers. As we move into this systems world, we aren't just making and selling boxes anymore. Increasingly, we are working with customers to design and redesign their basic business processes. In the future, Xerox won't just sell copiers. It will sell innovative approaches for performing work and for enhancing productivity.

But that means that our salespeople need to understand the customer's business, what the customer's real needs are, and how the customer is going to use our products. It's a partnership in which we take a more consultative approach.

Can you give an example?

Recently I was in Austin, Texas to make a call on a customer who until then had not wanted to have anything to do with us. The chief information officer at the company did not want any paper in his organization—and, therefore, he did not want any copiers because copiers generate paper.

When we finally got a meeting, our salesperson didn't say a word about copiers. He talked about product documentation. "Have you ever had any delays in getting your product out the door because your documentation wasn't ready?" Well, of course, the answer was yes. So our salesperson said, "We may have some capability to help you with that. We want to come in and look at the way you are doing documentation and see if we can figure out a better way to do it. If we can get your products out the door more rapidly, with more up-to-date documentation—and, by the way, more cheaply—would you be interested?"

Needless to say, he was. So our salesperson did a study to analyze how the company could organize its work more effectively. Not only that, he got the CIO a tour at another company down the road where we had already done a similar thing. And we made the sale.

You mentioned that the new business environment also means new competitors.

Our competitors are multiplying. Some of our Japanese rivals like Canon will certainly come along with us in this new digital world. But we will also be competing with companies that are now in the printing business or in the computer systems business. Increasingly, we are moving into the more complex business environment that a lot of computer companies already find themselves in—where we will compete against companies in some situations and cooperate with them in others.

What's more, we can't just say, "Out with the old and in with the new." Even as we develop new skills, define new ways of interacting with the customer, and confront new kinds of competition, we must still pursue our traditional business. We need to maintain and build on all the skills we have developed in the light-lens business. But at the same time, we must manage this technological transformation. We have to do two things at once.

So Xerox's new architecture is a direct response to these new business challenges.

Absolutely. Operating effectively in this more complex and volatile business environment requires the capacity to cope with change—and at a very rapid pace. The technology is changing quickly. The demands of the marketplace are also changing. What's more, they're both moving targets. They are going to continue to change. So we have to change the company itself.

On one level, that's simple: if we intend to sell customers our expertise as designers and implementers of new, more effective, and more productive business processes, then we had better make sure that our own organization is a showcase for better ways of working.

But even more important, we have to create a new organizational architecture flexible enough to adapt to change. We want an organization that can evolve, that can modify itself as technology, skills, competitors, and the entire business change.

What did Xerox's old architecture look like and why wasn't it adequate to these new challenges?

Until recently, Xerox had the structure, practices, and values of a classic big company. First, we were an extremely "functional" organization. If you were in manufacturing, you strived to make manufacturing as good as possible—and only secondarily to make the businesses that manufacturing affected work well. The same was true for sales, R&D, or any other function.

Second, we were staff-driven. Staff was primary, line businesses were secondary. I used to be the chief staff officer at Xerox, so I know what I'm talking about. At the time, in the mid-1980s, I viewed myself—quite inaccurately, I should add—as the number two person in the company. It certainly wasn't in terms of responsibility. There were people with billions of dollars worth of revenue and profit responsibility, whereas I didn't have any. And yet, I was the CEO's key adviser, and I ran the management process. I could wield a lot of weight without a lot of real responsibility.

That's how the power structure worked—and for good reason. The fact is, in a functional organization, you need many staff functions in order to make things work. To ensure that manufacturing, for example, was hooked to sales, both had to be hooked in to corporate.

What's wrong with a staff-driven functional organization in the new business environment that you describe?

It breeds dependence and passivity. In a functional organization, there is a natural tendency for conflicts to get kicked upstairs. People get too accustomed to sitting on their hands and waiting for a decision to come down from above. Well, sometimes the decision does come down. But sometimes it doesn't. And even when it does, often it comes too late, because market conditions have already changed or a more nimble competitor has gotten there first. Or maybe the decision is simply wrong—because the person making it is too far from the customer.

We used to go through a big exercise here every fall that went by the grandiose name of "resource optimization." Basically, the top-management team, aided by the marketing staff, took our total R&D spending—some $800 million to $900 million—and asked, "How are we going to allocate that budget?" First, we figured out how much money we could spend; then, we set priorities among our R&D projects against these resources to decide what got funded and what didn't.

It was as if the people who actually had to spend the money were living in a welfare state. They told us what they needed—"I need $50 million," "I need $40 million," "I need $30 million." Then they waited. At some point, somebody would come back to them and say, "What will happen if you only get $30 million instead of $40 million?" And then they would revise their plans according to the new figure. Finally, somebody would come back yet again and say, "OK, you get $35 million." That's how the process worked. It was completely crazy.

The old model is all too familiar. But what is the new model?

We are trying to break the bonds that tie up energy and commitment in a big company. Our goal is to make this $17 billion company more entrepreneurial, more innovative, and more responsive to the marketplace. In fact, we intend to create a company that combines the best of both worlds—the speed, flexibility, accountability, and creativity that come from being part of a small, highly focused organization, and the economies of scale, the access to resources, and the strategic vision that a large corporation can provide.

How are you going to accomplish that?

By redefining the three key components that make up any organizational architecture. The first, we call "hardware." These are the things managers usually think of as organizational structure, the formal processes by which we get things done: the business planning system, control mechanisms, measurement systems, reporting relationships, reward systems, and the like.

The second is "people"—the new skills managers need, including the kind of personality and character they must have to operate effectively in the new environment. Clearly, if you are running a functional organization, you go with functional experts and you manage with a lot of staff who hold the functions together. But in a decentralized, more entrepreneurial organization like we have now, you need a completely different kind of manager.

Finally, the third component of our new architecture is what we call "software." This is the most difficult to describe but probably the most important: the informal networks and practices linking people together, the value system, the culture. Any company that leaves these out of its organizational change effort is making a big mistake.

Why is that?

Think about how any organization works. The informal organization is often far more important than the shape of the organizational chart or who is in what box. Two companies that look exactly alike when you compare their organizational charts can turn out, in fact, to operate very differently.

A successful organization is one where the formal organization and the informal one work together, rather than always working against each other. When there isn't a good fit between them, too much human energy and creativity get wasted simply in making the organization work. But when you align the formal and the informal, most of your energy can be externally focused on achieving the objectives of the business.

Let's start with "hardware." What does Xerox's new formal organization look like?

When we began thinking through our new design two years ago, we needed to solve a basic paradox. For all the reasons I've mentioned, we wanted to get rid of our traditional functional organization and create more entrepreneurial business units.

And yet, for a company like Xerox, just setting up a bunch of small autonomous units isn't enough. Most of our businesses rest on a common technology base. What's more, we are dealing with a market where in many cases, the same customer is buying a variety of different products from us. So it's important not to lose the advantages of a situation where a salesperson can say, "I am the single person who will represent Xerox to you. And if you have a problem anywhere in the world, we will take care of you."

Take, for example, our total satisfaction guarantee. Anybody who buys a product from Xerox, no matter who sells it or where the customer buys it, has a satisfaction guarantee good around the world. That is very, very powerful. So how do you have speed, flexibility, autonomy and still be able to give the customer that guarantee?

How have you resolved that paradox?

We have created an organization that by its very design forces managers to confront—and manage—the necessary tensions between autonomy and integration. The centerpiece of our new architecture is a set of nine relatively independent business divisions. These are stand-alone businesses organized around specific products and markets and with profit and loss responsibility. Each division consists of a number of business teams, smaller entities tied to the marketplace by a specific customer need. In fact, the business team leader is the new entry-level general manager job in the company.

But these divisions don't exist in a vacuum. They are linked to other parts of the company whose responsibility is to leverage corporatewide resources and relationships in support of the divisions.

In a sense, we have turned the traditional vertically organized company on its side. At one end is technology, and we have retained an integrated corporate research and technology organization. At the other end is the customer. We have organized our sales and service people into three geographic customer-operations divisions, so that we can keep a common face to the customer. Between these two poles are the new business divisions. Their purpose is to create some "suction" on technology and pull it into the marketplace.

Finally, we have created a new unit called "strategic services" that will provide support to the business divisions in areas such as specialized manufacturing and purchasing, where economies of scale still make a big difference. And we've established a six-person corporate office, served by a much leaner corporate staff, to make sure all these units fit together in a strategically coherent way.

A big advantage of this structure is that it is remarkably easy to adapt. When we see new markets emerge or new technologies that don't fit into our current structure, we can simply add another business team or even a whole new division. Similarly, we can split a business division—or even eliminate one altogether—without changing the basic architecture of the company.

But as you point out, structure alone isn't enough to make a company work. How do you make sure that people manage the tension between autonomy and integration?

First, we are making it very clear to the presidents of these divisions that their primary responsibility is to make a success of these businesses. We in the corporate office will provide them with some constraints, some boundaries. If they feel they absolutely must cross those boundaries, then we expect them to tell us. But otherwise, they have the freedom to run the business, to make the objectives that they have communicated to us and to implement the strategy they have presented. Now of course, if we don't think their business plan is credible, then we'll object and talk with them about it. But the initiative is up to them.

Take the example of R&D spending I mentioned earlier. Before, setting priorities was a top-down process. Now, we're turning that process completely upside down. We're saying to the business division presidents, "You tell us what you are going to deliver and the resources you need to do it." They are setting the priorities now. And they should be because they are the people closest to the market.

See, they should be coming to corporate management the way an entrepreneur goes to a venture capitalist or banker. And we at the top should be looking at individual strategies and how those strategies come together for the company as a whole. We shouldn't be starting out with a fixed R&D budget. How the hell is anybody smart enough to say what the right amount should be?

But, of course, the business division presidents aren't really independent entrepreneurs. They're part of a big corporation.

That's true. Even as these division presidents develop their own businesses, they have to think with their corporate hat on as well. Again, technology provides a good example. One of the big challenges we face as we move into a technological environment founded on integrated systems is to create a common design architecture for our products. When all our products were stand-alone devices, a common architecture didn't much matter. The only goal was to optimize individual product design. As a result, we have copiers where the paper goes from left to right, and we have had printers where the paper goes from right to left.

But when it comes to designing an integrated system that includes all the copier components and all the printer components, obviously you can't have it both ways. So who decides what the unified architecture will be? I have no interest in making that decision alone. I am not anywhere near smart enough. And while we have created a technology architecture unit in our corporate research organization, I don't want our technologists deciding on their own either.

So, we have put a decision-making process in place that will directly involve the line managers of the business divisions. Of course, they will have to make compromises at times. There may well be an optimum technology for one division and a very different one for another. Since we can't afford to develop both, they will have to sit down and say, "What is the right technology for Xerox as a whole?"

So too at the customer end. We want the customer to integrate the company. We don't want to integrate the company and then go to the customer. The job of the customer operations divisions will be to build a relationship with the customer and then figure out with the business divisions how to integrate our offerings for the customer. To do this, the customer operations people will have to understand not only the customer but also the strategies and directions of the divisions.

It sounds like the tension between autonomy and integration is built in to the new managerial roles you have created. What kind of people do you need to perform those roles effectively?

We want people who can hold two things in their heads at the same time, who can think in terms of their individual organizations but also in terms of the company as a whole. Our architecture won't work if people take a narrow view of their jobs and don't work together.

To do that requires a total business focus—something that the narrow functional organization does not provide. This was a big issue when we were selecting our business division people. The fact is, we simply do not have as many general managers as we need who have that broad experience. There is no traditional career path inside Xerox to become a business team general manager or a business division president.

So how did you find the right people?

People may find this hard to believe, but we actually designed the entire organization without any particular individuals in mind. That goes for the key managerial positions in the business divisions. It even goes for my senior management team in the corporate office. We didn't discuss who would occupy those positions until we had decided what kind of structure we needed and wanted.

Once we had designed the new organization, however, we tried to be explicit about the type of people we need to operate within it. We started with a clean slate and defined the ideal characteristics of a manager in the new Xerox. We developed a whole new set of criteria for evaluating people—23 characteristics in all, 7 of which we decided were absolutely critical for running a business division or team. Many are things that, frankly, we never thought of as so important before.

Like what?

Take strategic thinking. In the old organization, that wasn't very critical for most people other than those at the very top of the company. But if you're a business division president in the new architecture, it's crucial.

Another example is strategic implementation. We have a lot of people who were strategists or planners but who never had any responsibility for implementing their plans. But knowing how to think strategically is no longer good enough. You must also know how to act strategically.

And then, clearly, the "software" items are critical: teamwork, the ability to delegate and to empower subordinates. Other criteria have more to do with personal character. For instance, one we call "personal consistency." That's a fancy way of saying integrity, personal courage, "grace under pressure." We don't want people in these senior management positions who are all over the map, or who are constantly holding their fingers up in the air to see which way the wind is blowing.

What did you do with these characteristics once you identified them?

I sat down with a few top managers, and we rated some 45 people we thought were candidates for roughly the top 25 jobs in the company, including the top position in all the business and customer operations divisions. First, each of us rated everybody individually. Then we compared and discussed our individual ratings of each candidate and agreed on a common rating. And we did it without regard to particular jobs. It was only afterward that we considered individuals for specific positions.

We selected some very nontraditional people based on this approach. People came to the surface who wouldn't have in the old system. For instance, 3 of our 9 new business division presidents have been with Xerox for less than a year. And about one-third of the top 40 people in the company have been at Xerox for less than three years.

What's more, we continued this rigorous selection process down one more level into the organization. Recently, the 12 business division and customer operations presidents used it to choose their business team general managers and support staff. They rated more than 100 people for about 50 jobs.

Assuming you've found the right people, how are you managing for the new behaviors you need?

One of the things I learned long ago is that if you talk about change and then leave the reward and recognition system exactly the same, nothing changes. And for good reason: people quite rationally say, "I hear what he is saying, but it's not what I get paid to do or what I get promoted for. So what's in it for me?" Therefore, if you are trying to change the way you run a company, one of the most visible things you have to change is the way you compensate, the way you reward and recognize people.

The compensation of senior executives has come in for a lot of criticism recently. Did that influence how you designed your new reward system?

There is a lot of emotion around the issue of executive compensation right now. But frankly, when we began to design our new compensation system, we were focused almost exclusively on our internal goals: to encourage the new kinds of behavior essential to making the new organizational architecture work and to achieving our business goals. As we developed our plan, we discovered that it also provides exactly what shareholders want. It aligns individual compensation with the strategic objectives of the company. It rewards good performance and penalizes poor performance.

How does the new plan work?

Like most companies, we had a long-term incentive plan based on stock options and a bonus. But the way it was structured, you were guaranteed a payout unless the company completely fell apart. On the other hand, if the company did extremely well, the payout didn't really go up a lot. So the plan didn't function effectively as an incentive for people to work together and aggressively drive the company forward.

We decided to add two new features to our incentive plan for the people in roughly the top 50 jobs in the company. First, while there continue to be stock options, we also have a program for awarding people what we call "incentive shares," outright grants of Xerox stock. However, these shares are earned only if the company meets certain hurdles fixed in advance and based on return on assets. Everybody has to work together to make those numbers, and the numbers don't get reset from year to year. If we miss it the first year, that means that next year's target is all that much tougher. We have to do better in 1992 than we did in 1991 and improve again in 1993 and in 1994.

What's more, we believe that to be fully committed to the long-term success of the company, you have to have some skin in it yourself. Therefore, we have made it a condition of participation in this plan that senior managers must actually purchase one year's salary worth of Xerox stock.

Compared with the old plan, there is potential for bigger payout, but there is also a lot more at risk. There is more risk because you have to buy stock. Second, if we do not make our return-on-assets targets, we receive no incentive shares at all. So now we have a situation where people could actually lose money if the company doesn't perform. But if we perform well, we can all do much better.

But that long-term incentive plan only affects the top people in the company. How are you changing reward systems for people further down in the organization?

We have also made important changes in our bonus plan, which covers the top 2,000 people in the company. In the old world, a bonus was "additive." Say, 30% was based on how the corporation did during the year. Maybe 50% was based on how your division did. And 20% was based on how you as an individual did.

But the effect of such a system is to encourage people to say, "Well, I don't have a big impact on the company, so I'll just optimize my individual objectives or my division's objectives. And whatever happens in the corporation will just be icing on the cake."

In our new reward system, we have what I call a "multiplicative" approach. The categories are the same—corporate performance, division performance, personal performance. But instead of adding them together, we multiply them against each other. So if a manager does well personally—say 120 on a scale of 100—but the company as a whole does poorly—say 70 on a scale of 100—his bonus will suffer. He will get only 84% of his intended bonus.

On the other hand, if a manager does well and through her efforts and those of many others the company also does well—say both score 120 on a scale of 100—then her bonus increases substantially to 144% of the intended bonus.

This changes the whole dynamic. The message to the individual manager—not only in terms of his or her self-interest but in terms of pressure from peers—is "you had better be contributing to the company as a whole."

Finally, we have set aside an extra pool of money—up to an additional 10% of the entire bonus pool—for the managers who show exemplary performance in these soft values I've been emphasizing. I have the option of rewarding these managers handsomely.

Why are you placing so much emphasis on the intangibles of "software"?

Because the hardest stuff is the soft stuff—values, personal style, ways of interacting. We are trying to change the total culture of the company. When you talk about that in general terms, everybody is all for it. But when you talk about it in terms of individuals, it is much tougher. And yet, if individuals don't change, nothing changes.

What holds people back?

Imagine someone who is 45 years old, who has been a manager for 20 odd years and a successful one, who has gotten to, say, the vice president level. All of a sudden, we are saying to that person, "We know you've been successful, but we're changing the rules. You have to do things differently." That triggers enormous insecurity. Because it is perfectly natural for the person to be thinking, "I am in this nice corner office because of the way I operate. And yet I hear all this stuff about change."

Maybe the individual even understands the need for change—intellectually. But the real problem is, "How am I going to act when I get in trouble? Am I really going to trust all these new approaches? Am I really going to be comfortable with delegating responsibility and authority? Am I really going to take full accountability and not cleverly spread it around so no one can throw stones at me?"

We are asking people to take initiative, make decisions, and stand by them. Well, that's great when things go well. But what about when they don't?

What do you say to somebody like that?

You just have to be as explicit as you can with everyone about the new criteria for success and then give people honest feedback about where they stand. For instance, when we rated managers against our new job selection criteria, we also told them, "Here are the three or four characteristics where you are weakest. It is important that you work on these weaknesses because part of the way we will evaluate your success is your ability to work in this new manner."

How have people reacted?

It's been something of a shock. In our old performance appraisal system, we had five levels—and good performers regularly scored fours or fives. But with the new system, we rate people on a seven-point scale. Because we were using this for selection purposes, we tried to be completely honest. So, many people are getting threes and twos on a specific characteristic. They look at it and say, "Wait a minute. I'm one of the people being selected?"

Now ideally, we would like to have a portfolio of players in these jobs who are scoring fives, sixes, and sevens. But realistically, we simply don't have many of those people yet. In the new system, the highest possible score is 161. Well, our top people are in the 50s, and the bottom people are in the 20s. Believe me, that has gotten people's attention.

But we're not doing it just to be perverse. The change process we have begun is extremely difficult. It's hard for any organization to get out of its history. We have to be very diligent so that old habits, practices, and behaviors don't sneak back in.

How common is that?

My favorite example comes from the management team that designed our new architecture. Here is a group of 20 people who, more than anybody in the company, have internalized the new approach to running things (see "Architecting a New Organization"). And yet, when it came to deciding how to organize the business divisions, this group recommended that we start with the business division staff—the head of personnel, head of finance, etc.—and that the staff help each division president choose his or her key operating people. In other words, they immediately fell back into the traditional top-down, functional approach.

I said, "Hold on. Don't we have it backwards? I thought we were trying to run this business by the line? Why would you select the staff people to help the presidents select the line? Why don't we do it just the reverse? Who are the key line managers required to run these businesses? We can help them figure that out. They don't need to hire individual staff people to do that. We can help them put their team leaders in place, then the presidents and their team leaders can decide together what kind of staff support they need."

At first, they were convinced I was wrong. But as we talked it through, you could see a shock of recognition go through the group. Everyone started to say, "Well, I think we just violated one of the principles with this recommendation." They began to realize that it would send exactly the wrong message to the organization. Finally, somebody said to the person who had made the original proposal, "I think you'd better back off on this. Not only is he the boss, but I think he is right on this one."

I keep reminding people that what we are doing isn't easy. In fact, if they are finding it easy, then they should be concerned because that's a sure sign they are not really succeeding at it!

Architecting a New Organization

When Paul Allaire became CEO of Xerox in August 1990, he knew he wanted to re-examine the fundamental operating principles of the company, but he was unsure of the best way to do it. So that October, Allaire appointed a team of six young Xerox managers to examine what kind of structure and practices the company needed to be successful. The group, which Allaire called the Future Architecture team, kicked off a systematic 15-month design process that would involve more than 75 Xerox managers from throughout the organization. This participatory design process did not just define the principles of Xerox's new architecture. It also became an effective mechanism for internalizing those principles and enacting them in the senior management ranks.

None of the managers chosen for the Future Architecture team was an expert in organizational design. But all had successful track records as line managers in jobs three and four levels down in the organization. They also had reputations for outspokenness, with strong ideas about what was right and wrong with Xerox.

The group, which became known as the "Futuretecture" team, had a broad charter: to reflect critically on the basic operating premises governing the company; to educate themselves about the latest thinking in strategy, management, and organizational design; and, most important, to go outside the company in search of new organizational models.

Over the next three months, the futuretects developed four possible scenarios for Xerox's future, which they presented to senior management in February 1991. The first was simply to make Xerox's traditional functional organization work better. The second was to divide the rest of the company according to key market segments. The third alternative was to create a "transnational" company organized by geographic region.

The futuretects strongly preferred a fourth scenario: to establish independent global business divisions. But their recommendation came with a warning. The formal structure chosen was less important than the informal behaviors, values, and attitudes that guided people within it.

Over the summer, some of the futuretects worked with Allaire and his staff to flesh out the business division concept and define the particular divisions that Xerox would need. Then in late September 1991, Allaire chartered another informal team, known as the Organizational Transition Board, to work out the details of the new model.

The OTB consisted of 15 senior managers and 5 support staff drawn from the Futuretecture team and Allaire's staff. The theory behind this grouping was to bring together a broad cross section of the most influential senior managers in the company, then give them a relatively small staff to force them to work through the new principles on their own. The goal of the OTB was not just to redesign the company but also to create an understanding among key senior managers about where the company was going.

For the next few months, the OTB met for a minimum of two full days, every other week. Defining the new organizational structure was the easy part. Addressing the hard-to-define "software" and "people" issues was more difficult. What was the ideal relationship between the business divisions and customer operations divisions? What was the job of the new corporate office at the top of the company? And what kind of new skills did managers need to function in the new environment? The OTB set up subgroups, which involved another 50 people from throughout the company, to address these and other issues.

According to Allaire, the OTB has turned out to be "far more important than any of us could have imagined. These people hammered out these principles without knowing where they would end up in the new organization they were creating. Through that process, they internalized those principles to a remarkable degree. We in the corporate office often found ourselves behind them."

The OTB's final recommendations were presented to Xerox's annual senior management meeting last February. Many of the people who served on the Future Architecture team and the OTB are now in key leadership positions in Xerox's new organization. Having redesigned the company, they are now designing their own organizations within it.

Everybody knows about the benchmarking tradition at Xerox. Is there any company that you benchmark the entire organization against? Are there companies operating today as you want Xerox to be operating, say, five years from now?

Unfortunately, no. I say "unfortunately" because it would be a lot easier if there were. We are quite shameless in our willingness to steal good ideas from elsewhere. But there aren't any companies that operate as a company the way we intend to.

The important thing to remember is that this new organizational architecture is just the beginning. Our ultimate goal is to organize the entire company into self-managed work teams—or what we call "productive work communities."

I envision a time when this company will consist of many, many small groups of people who have the technical expertise and the business knowledge and the information tools they need to design their own work process and to improve and adapt that process continuously as business conditions change. These work groups will be tied directly to the customer. They will be working in a much less supervised environment. And they will have the resources they need to redesign their work environment and modify their behaviors as needed to achieve their objectives in the marketplace (see "From Strategy to Product: The Story of PaperWorks").

From Strategy to Product: The Story of PaperWorks

Xerox has redefined its strategy and redesigned its organizational architecture. But how is the company trying to translate the ideals of strategy and structure into the reality of new products and markets? One example is the story of a new Xerox software product called PaperWorks.

PaperWorks is a $250 software package that allows users to access their personal computers from any fax machine in the world. Using special paper forms, an individual can instruct his computer system to store, retrieve, and distribute files without having to be at the computer itself. For example, a business traveler might use PaperWorks to instruct her PC to forward any faxes that arrive at her home office to her hotel fax machine. Or she could send a copy of a magazine article by fax, store an image of the article in the computer, and then have it fax copies to the people on a predetermined distribution list.

PaperWorks is a concrete example of Xerox's strategic intent to bridge the gap between paper-based and electronic information. But perhaps even more important, the way PaperWorks moved from intriguing idea to actual product is a good illustration of the more entrepreneurial and innovative work environment that the company is trying to create.

The idea for PaperWorks first took shape in the late 1980s at Xerox's Palo Alto Research Center. PARC research on advanced image processing had given birth to a variety of new technologies, including one known as "glyphs"—a technique similar to bar coding that uses small marks to encode information on paper that can be read by computer. Meanwhile, more than a decade of field work by PARC anthropologists had generated a fundamental insight about work: for all the loose talk about the "paperless office," paper remains a ubiquitous and extremely "user friendly" way for people to present and use information. These two streams of research led PARC researchers to explore ways to use paper as a new kind of computer "interface."

One of those researchers was Walt Johnson, a cognitive psychologist in PARC's System Sciences lab. In 1988, Johnson and his colleagues began experimenting with glyphs to create paper forms for the remote retrieval of documents. In the summer of 1990, Dennis Andrews, the head of Xerox's computer software division, X-Soft, learned of this work and asked product-development manager Jennifer Ware to visit PARC and look at the technology with an eye toward developing a product concept. "Dennis told me, 'Think PCs, fax machines, and software. Then let me know what you want to do,'" Ware remembers. Andrews also had one more condition: the new product had to be ready in a year.

To meet that tight deadline meant creating a small and highly collaborative product-development team spanning research, development, and marketing. In addition to Johnson and Ware, Xerox marketing manager, Carolyn Grossman, moved to PARC to join the PaperWorks team. For eight months before the project was approved by Xerox senior management, Andrews supported the team with miscellaneous funds culled from the X-Soft budget. And because the PaperWorks team was based at PARC, members could draw on the wealth of expertise of PARC researchers—including those who were not officially part of the project.

Once formal approval came in January 1991, the PaperWorks team also worked closely with potential customers. One was the New York office of UNICEF. At first, UNICEF staff used PaperWorks mainly to speed the flow of incoming faxes around the 12 floors of their Manhattan office building. But they soon realized the new software could help them coordinate the work of UNICEF's far-flung global network of health researchers, field workers, and government experts.

For example, every week the coordinator of UNICEF's Guinea Worm Eradication Project faxes a weekly information bulletin to the roughly 40 experts involved in the project. Most are in West Africa, but others are at universities and government research institutes around the world. Before PaperWorks, she would start faxing around 10:00 P.M. and, because the quality of most phone lines to Africa is so poor, spend about 6 hours making sure every fax got through. With PaperWorks, she faxes a form to her computer instructing the system to fax the bulletin, stored on the computer's hard disk, to everyone on the project distribution list. The computer works through the night until all the faxes are successfully delivered.

By the fall of 1991, the PaperWorks team was well on its way to finalizing its product. As it took shape, however, Xerox corporate managers began to realize the potential for a whole family of products based on the same underlying PARC technologies. So even as they rushed to meet their January 1992 deadline, the PaperWorks team became involved in a much bigger development project. During a month of 12-hour days and 7-day weeks, team members helped senior managers write the business plan for Xerox's new Advanced Office Document Services (AODS) division.

The new division was announced with the rest of Xerox's new architecture last February. One month later, X-Soft released the first version of PaperWorks—15 months after the official start of the product-development project.

Of course, whether PaperWorks will succeed in the marketplace remains an open question. But as an example of what the company calls "productive work communities," the PaperWorks product-development team may well be a kind of model for the company's future.

"This project is unlike anything I've seen in my 15 years at Xerox," says Jennifer Ware. "The team made all the key decisions together, with little senior management review. That gave us the ability to act very quickly."

Now that Xerox's new architecture is in place, there is even a formal process for facilitating the kind of rapid new-product development that happened informally with PaperWorks. An office for "advanced technology and market development," located in Xerox's corporate research and technology organization, is responsible for helping the new business divisions and the company as a whole identify new technologies and markets and develop business plans for exploiting them.

Put another way, Xerox's new architecture won't really be complete until what you are now doing at the top of the company is also taking place at the lowest levels—until everybody becomes an "organizational architect."

Right. Now, a lot of companies are experimenting with creating self-managed organizations. You can see a factory that operates that way or maybe a warehouse or service facility somewhere. Indeed, you can find pockets of organizational innovation at a lot of big companies. I think we've visited just about all of them.

But it doesn't provide the answer that we are looking for: How do you do it on the scale of an entire large corporation? There is no big company that is managed that way. There is no example—or even a very good theory—of how to run a multibillion dollar company that way.

We intend to keep looking and keep experimenting. Because I firmly believe that is the future. And as an organization, we are committed to living in that future.

PART

III

Managing New Psychological Frontiers

1
The New Boundaries of the "Boundaryless" Company

Larry Hirschhorn and Thomas Gilmore

In an economy founded on innovation and change, one of the premier challenges of management is to design more flexible organizations. Companies are replacing vertical hierarchies with horizontal networks; linking together traditional functions through interfunctional teams; and forming strategic alliances with suppliers, customers, and even competitors. Managers are insisting that every employee understand and adhere to the company's strategic mission without distinction of title, function, or task.

For many executives, a single metaphor has come to embody this managerial challenge and to capture the kind of organization they want to create: the "corporation without boundaries." General Electric CEO Jack Welch has eloquently described this new organizational model. "Our dream for the 1990s," Welch wrote in GE's 1990 annual report, "is a boundaryless company . . . where we knock down the walls that separate us from each other on the inside and from our key constituencies on the outside." In Welch's vision, such a company would remove barriers among traditional functions, "recognize no distinctions" between domestic and foreign operations, and "ignore or erase group labels such as 'management,' 'salaried,' or 'hourly,' which get in the way of people working together."

Managers are right to break down the boundaries that make organizations rigid and unresponsive. But they are wrong if they think that doing so eliminates the need for boundaries altogether. Indeed, once traditional boundaries of hierarchy, function, and geography disappear, a new set of boundaries becomes important.

These new boundaries are more psychological than organizational.

They aren't drawn on a company's organizational chart but in the minds of its managers and employees. And instead of being reflected in a company's structure, they must be "enacted" over and over again in a manager's relationships with bosses, subordinates, and peers.

Because these new boundaries are so different from the traditional kind, they tend to be invisible to most managers. Yet knowing how to recognize these new boundaries and use them productively is the essence of management in the flexible organization. And managers can find help in doing so from an unexpected place: their own gut feelings about work and the people with whom they do it.

The Challenges of Flexible Work

In the traditional company, boundaries were "hardwired" into the very structure of the organization. The hierarchy of occupational titles made manifest differences in power and authority. Independent functional departments coordinated pools of specialized expertise. Dedicated business units were a reflection of a company's products and markets.

This organizational structure was rigid, but it had a singular advantage: the roles of managers and employees within this structure were simple, clear, and relatively stable. Company boundaries functioned like markers on a map. By making clear who reported to whom and who was responsible for what, boundaries oriented and coordinated individual behavior and harnessed it to the purposes of the company as a whole.

The problem is that this traditional organizational map describes a world that no longer exists. New technologies, fast-changing markets, and global competition are revolutionizing business relationships. As companies blur their traditional boundaries to respond to this more fluid business environment, the roles that people play at work and the tasks they perform become correspondingly blurred and ambiguous.

However, just because work roles are no longer defined by the formal organizational structure doesn't mean that differences in authority, skill, talent, and perspective simply disappear. Rather, these differences present both managers and employees with an added challenge. Everyone in a company now must figure out what kind of roles they need to play and what kind of relationships they need to maintain in order to use those differences effectively in productive work.

Take the simple example of an engineer on an interfunctional product design team. To be an effective participant on the team, the engineer must play a bewildering variety of roles. Sometimes she acts as a technical specialist to assess the integrity of the team's product design; at other times she acts as a representative of the engineering department to make sure that engineering does not get saddled with too much responsibility while receiving too few resources; then again, in other situations she may act as a loyal team member to champion the team's work with her engineering colleagues.

No one role exhausts the kinds of relationships she must engage in to make the team work. The engineer will probably play all three roles at least once while she's on the team. But how does she know which role to play when? And how can she be sure that the rest of the team knows which role she is playing at any particular moment in time?

In the corporation without boundaries, then, creating the right kind of relationships at the right time is the key to productivity, innovation, and effectiveness. But good working relationships don't happen automatically; they are not the simple product of good feelings, team spirit, or hard work. In fact, opportunities for confusion and conflict abound in a flexible organization.

Imagine the following typical interaction between a shop-floor worker and an engineer at a company trying to create a team environment. The worker takes the company's commitment to teamwork seriously and, in an attempt to learn how and why product engineers make the decisions they do, asks an engineer to explain the criteria he used to approve some design changes on a blueprint.

The worker has focused on the task. He wants to be the engineer's colleague. Perhaps he imagines that together they can develop a new and more productive way to divide up the work. Unfortunately, the engineer hears the question not as a simple request for information but as an implicit attack on his authority. So he answers vaguely and dismissively, making it clear he doesn't think much of the worker's question.

The worker feels put down and doesn't press his question. But instead of trying to understand why the engineer reacted the way he did, the worker simply chalks up the response to the contempt that "elitist" engineers feel for "uneducated," blue-collar workers.

The engineer and worker don't know how to manage the psychological boundaries that order their relationship. During their interaction, they draw on a succession of distinctions—between expert and novice, superior and subordinate, exploiter and victim. However, because neither has an accurate "map" to figure out the kind of relationship they are in and what boundary they have encountered, the interaction that was intended to make them more effective colleagues only serves to separate them. The result is a failed encounter and an unproductive relationship.

In fact, too much focus on eliminating old boundaries can cause managers to misunderstand their fundamental role in the flexible organization. All too often, managers think that getting rid of boundaries also means doing away with conflict. They assume that once the company breaks down the walls that "get in the way of people working together," employees like the engineer and the worker will put aside what divides them and unite behind the company's mission. Differences of authority, talent, or perspective will no longer be a source of friction.

Nothing could be further from the truth. As traditional boundaries disappear, establishing such differences becomes simultaneously more important and more difficult. Flexibility depends on maintaining a creative tension among widely different but complementary skills and points of view. In the theater, a demanding director can elicit an especially brilliant performance from an actor. Similarly, an accomplished actor can help a director better understand his own vision of a play. So too in the workplace, where demanding subordinates can make for better bosses—and a brilliant marketing department can push manufacturing to perform at its best.

But this kind of creative tension does not come easily. As the tasks, roles, and outcomes of work become more uncertain, clashes of opinion and perspectives become more likely. Because they may signal that a work group is approaching a boundary that needs managing, such conflicts can be healthy and productive—*if* they are contained or bounded so they don't become overwhelming.

Therefore, managers in flexible organizations must focus on boundary management. They must teach people what new boundaries matter most, then how to recognize such boundaries in their relationships with others. Finally, good boundary managers encourage employees to *enact* the right kinds of boundaries at the right time, as a director helps talented actors take up and perform the roles of a good play.

Remapping Organizational Boundaries

What psychological boundaries must managers pay attention to in flexible organizations? We call them the "authority" boundary, the "task" boundary, the "political" boundary, and the "identity" boundary. Each is rooted in one of four dimensions common to all work experiences. At the same time, each poses a qualitatively new set of managerial challenges in the new work environment. And each boundary can be recognized by the characteristic feelings it evokes. If managers are attentive, they can use these feelings as clues to assess whether their relationships at the boundary are working effectively. (See Exhibit I.)

THE AUTHORITY BOUNDARY. Even in the most boundaryless company, some people lead and others follow, some provide direction while others have responsibility for execution. When managers and employees take up these roles and act as superiors and subordinates, they meet at the authority boundary.

The authority boundary poses the question: "Who is in charge of what?" In most companies, that question used to be relatively easy to answer. Those in authority were easy to identify. Bosses issued orders, and workers followed them. Management was primarily a matter of effective monitoring and control.

But in more flexible organizations, issuing and following orders is no longer good enough. The individual with the formal authority is not necessarily the one with the most up-to-date information about a business problem or customer need. A manager may lead a quality team, for example, that includes not only her peers but also her boss. Or an account rep may ask his boss to join the account team at a critical phase in the work with an important customer.

In such situations, subordinates face the far more complicated task of adequately informing their superiors and helping them to think clearly and rationally, even as they work to implement their superiors' requests. Paradoxically, being an effective follower often means that subordinates have to challenge their superiors. After all, allowing superiors to act foolishly only undermines them.

What it takes to be an effective superior is similarly complicated. Managers need to take charge and to provide strong leadership. But in the process, they must also remain open, even vulnerable, to criticism and feedback from below. If subordinates need to challenge in order to follow, superiors must listen in order to lead.

When superiors and subordinates work well together, both can play their respective roles. Subordinates feel trusted by their superiors, and that feeling of trust frees them up to exercise initiative at work. Supe-

Exhibit I.

A Manager's Guide to the Boundaries That Matter

Key Questions		Necessary Tensions	Characteristic Feelings +	−
"Who is in charge of what?"	AUTHORITY BOUNDARY	How to lead but remain open to criticism. How to follow but still challenge superiors.	trustful open	rigid rebellious passive
"Who does what?"	TASK BOUNDARY	How to depend on others you don't control. How to specialize yet understand other people's jobs.	confident competent proud	anxious incompetent ashamed
"What's in it for us?"	POLITICAL BOUNDARY	How to defend one's interests without undermining the organization. How to differentiate between win-win and win-lose situations.	empowered treated fairly	powerless exploited
"Who is—and isn't—'us'?"	IDENTITY BOUNDARY	How to feel pride without devaluing others. How to remain loyal without underming outsiders.	proud loyal tolerant	distrusting contemptu- ous

riors feel simultaneously supported and challenged by their staffs, which allows them to lead.

But when people don't work effectively at the authority boundary, other feelings predominate. Subordinates who don't believe that their bosses trust them can become either rebellious or excessively dependent and cautious—opposite symptoms that reflect the same underlying problem. Similarly, superiors who are not challenged by their employees may feel invulnerable, as if they "can do no wrong." At the same time, the lack of support from their subordinates may make them suspicious and overcontrolling.

THE TASK BOUNDARY. Work in complex organizations requires a highly specialized division of labor. Yet the more specialized work becomes, the harder it is to give people a sense of a common mission. This contradiction between specialized tasks and the need for shared purpose helps explain why teams have become such a popular form of work organization in recent years. Teams provide a mechanism for bringing together people with different but complementary skills and tying them to a single goal: designing and manufacturing a new product, say, or providing integrated service to an important customer.

But in order for teams to work, those involved must manage their relationships at the task boundary. Here the critical question is, "Who does what?" People in task relationships divide up the work they share and then coordinate their separate efforts so that the resulting product or service has integrity.

In the traditional organization, managing task relationships was largely a matter of overseeing the formal interactions among R&D, manufacturing, marketing, and the other classic functions. But in the new team environment, people from all of these areas are mixed together. Increasingly, individuals have to depend on others who have skills and resources they cannot control and often don't even understand. To be effective, they cannot simply ignore the work of others— in effect, to say "it's not my job"—any more than a subordinate can simply follow the orders of his or her boss. Indeed, their own performance may depend directly on what their colleagues do. So, while focusing primarily on their own task, they must also take a lively interest in the challenges and problems facing others who contribute in different ways to the final product or service.

When task relationships with coworkers go well, people feel proud of their work, comfortable about their dependence on others, and confident that they have the resources and the skills necessary to get the job done. But when a work group has problems defining the task, dividing up responsibilities, and apportioning resources, individual members begin to feel incompetent, unable to accomplish their work, and sometimes even ashamed of the job they've done.

THE POLITICAL BOUNDARY. In most companies, "politics" is a term of pure derision. Indeed, one of the promises of the corporation without boundaries is to make the company into "one big happy family" and eliminate politics from the workplace once and for all. But this promise contains within it a potentially dangerous mistake.

Politics involves the interaction of groups with different interests, and any large complex organization contains many such groups. R&D has a legitimate interest in long-term research, manufacturing in the producibility of a product, marketing in customer acceptance. A union member who confronts a foreman over an alleged contract violation, a regional vice president who wants to make sure her factories get more investment funds, and the director of a research lab who tries to protect his scientists from intrusions from marketing are all engaged in necessarily political relationships.

These relationships can be extremely useful to senior managers, because they mobilize the different interests and perspectives that together add up to a comprehensive view of the entire situation. Political activity becomes detrimental only when people are unable to negotiate and bargain in productive ways and when they can't define their interests broadly enough to discover mutually beneficial solutions.

When managers meet at the political boundary, they view one another as members of distinct interest groups with different needs and goals. They pose the question, "What's in it for us?" Then, by negotiating and bargaining with each other, they form coalitions to further their ends and develop strategies and tactics for advancing their interests. At the political boundary, people face the challenge of defending their interests without undermining the effectiveness and coherence of the organization as a whole. They must try to distinguish between "win-lose" and "win-win" strategies.

When groups in a company do this effectively, people tend to feel powerful. Staff members believe they are treated fairly and rewarded adequately. But when political relationships go badly, members of a particular work group can feel unrecognized, underrepresented in important decisions, and exploited.

THE IDENTITY BOUNDARY.   The corporation without boundaries seems to offer employees a common identity, the kind that Jack Welch suggests when he talks about erasing the "group labels . . . which get in the way of people working together." In fact, people have a multitude of group identities at work. Sometimes these identities are a product of a particular occupational or professional culture: attorneys, engineers, software programmers, even shop-floor workers. Sometimes they are rooted in the local work group: the team, department, or regional office. And sometimes their origins are more personal, grounded in the individual's experience as a member of a particular race, gender, or nationality.

The distinguishing characteristic of such relationships is the group's "sameness." When people begin to think in terms of "us" versus "them," of their in-group as opposed to other out-groups, they are engaged in a relationship at the identity boundary. Unlike the political boundary, which is about interests, the identity boundary is about values. Put another way, the identity boundary raises the question, "Who is—and isn't—'us'?"

People acting at the identity boundary trust insiders but are wary of outsiders. They seek out people who seem like themselves and take for granted the value of their own group perspective. The members of a local office may feel that headquarters has no understanding of how their region really operates. Scientists in an R&D lab may feel that marketing people have no conception of what makes them tick. Women managers may be convinced that their male colleagues don't respect their distinctive style of managing. Or foreign nationals in a multinational company may believe that headquarters overseas cannot really grasp the subtleties of the local market.

Identity relationships are important because they tend to be extremely energizing and motivating. In a workplace where effective performance increasingly depends on employee commitment to and engagement in the job, organizations need to tap this energy source and put it to productive use. That's why companies like Xerox, Corning, and Levi Strauss have encouraged diversity at all levels of the organization.

But relationships at the identity boundary also run the risk of disrupting the broader allegiances necessary to work together. For this reason, creating and supporting a sense of élan or team spirit—"we are the best group"—without devaluing the potential contribution of other groups is the real challenge of work at the identity boundary.

When organizations strike this balance, people feel loyal to their own groups and also maintain a healthy respect for others. But when team spirit is accompanied by contempt for others who don't share the same values or experience, identity relationships become extremely disruptive.

It's important to remember that these four psychological boundaries don't exist in isolation from one another. In any work experience, they interact dynamically. Consider the example of the unsuccessful product development team in "Decoding Boundary Mistakes: The Team That Failed."

Decoding Boundary Mistakes: The Team That Failed

Senior managers at a midsize office equipment manufacturer faced a serious threat. New competitors were introducing lower priced products that outperformed the company's traditional product line. Managers knew they had to counter with a new product of their own. But they worried that their company, which was organized along strong functional lines, simply could not respond fast enough.

For the company's chief operating officer, the solution was to create a prototype team that would design a new, more advanced version of the company's main product. The team approach, the COO argued, could benefit the company in two ways. It would be the fastest means of bringing to market a product equal to the changing competitive situation. Even more important, the team would be a laboratory for organizational learning in which the company could experiment with a more flexible work organization. Despite the skepticism of some of the company's functional vice presidents, the CEO agreed.

The COO appointed representatives from marketing, manufacturing, engineering, and finance to a 12-member product redesign task force. Their mission, he explained, was not only to innovate a new product but also to invent a whole new way of working together. For example, the team would be entirely "self-managed"; members would select their own leader from among themselves. To support the team in its efforts, the COO hired a trainer to teach members the new skills—brainstorming, problem solving, and group dynamics—necessary to work effectively in a team environment.

At first, task force members shared the skepticism about teams that they picked up from their superiors. But as they developed their team skills— how to define problems systematically, to give everyone a hearing, and to reach consensus—their skepticism melted away. As the COO watched the team grow, he felt extremely encouraged. The old fighting was gone. He was convinced that he'd fashioned the first successful interfunctional group in what had always been a tradition-bound and turf-conscious company.

The task force developed a new product design right on schedule and presented it to the company's executive committee at a special meeting. In a gesture that emphasized the group's team spirit, the lowest status member of the task force, a manager from purchasing, directed the presentation. Everything proceeded without a hitch.

However, when it came time to discuss the team's proposals, the meeting broke down. Company executives, naturally enough, asked tough questions to test the task force's design concept. Was this product really an improvement on current company lines? Wouldn't it just confuse the customer? What more could the team do to bring down prices?

But instead of responding to these legitimate questions, the team members kept defending their original proposal. They didn't think *any* aspect of their design could be modified. And team members seemed ready—too ready—to leap to the defense of fellow team members from different functions. When one senior manager, for example, argued that the product's manufacturing costs were too high to be competitive in the current marketplace, it was a team member from *marketing* who insisted that the costs couldn't be lowered. Similarly, when another senior manager criticized the product design as too complex and therefore difficult to manufacture, it was someone from *manufacturing* who defended the complexity as necessary for producing a first-class product.

The more the senior managers pushed, the more the team members dug in their heels. But this cycle only succeeded in convincing top management that the task force had become inflexible. The executive group rejected the product design and asked the team to go back to the drawing board. When the team members left the room, several senior managers insisted that the COO find new people for the team.

Why did this office equipment company's attempt at interfunctional product design go wrong? The answer is that senior managers didn't know how to organize the team's boundaries so that it could work effectively. In the process of bringing the team together, at least three boundary mistakes occurred.

1. The COO did not create a clear authority boundary. On the one hand, he gave team members a grandiose sense of themselves and their mission. They were to be the new model for the entire company. On the other hand, once he established the team, he refused to play a leadership role. Because the team was self-managed, no one had the authority to make hard trade-offs among conflicting goals.

2. The team coped with the absence of authority by enacting an identity boundary that was too strong. Once told that they were pioneers in interfunctional work, team members became convinced they were the potential saviors of the company and developed a sense of themselves as separate from everyone else. They were to be "different," "new," an "elite group."

3. An overly strong identity boundary prevented team members from creating the internal task and political boundaries they needed to do their work. The price for sustaining the sense of difference *between* the team and the rest of the company was the suppression of important differences *among* team members.

For example, the team never found a way to create the appropriate task boundaries so that each member used his or her special expertise on the design project for the good of the whole. The manufacturing engineer didn't sufficiently challenge the design engineer's work from the perspective of producibility. The marketing representative rarely questioned the engineer's cost assumptions from the perspective of customer acceptance and pricing. As a result, the team was never able to take advantage of the different skills team members had or to optimize them in the product design.

In addition, because task force participants were so committed to their identity as members of an elite team, they lost sight of the politics of product design. Yet for the team to function effectively, the managers *had* to think of themselves as something more than team members. They were also political representatives of important interest groups throughout the company. Because they failed to represent these groups—for example, manufacturing's interest in product simplicity or marketing's interest in a product that could be distributed through established channels—they could not effectively sell the new design to their own colleagues.

The unfortunate result: a product design that couldn't be defended on either technical or political grounds and renewed skepticism throughout the company about the usefulness of teams.

A chief operating officer establishes an interfunctional team to design a new product. He believes that in order for the team to work, it is enough to encourage group cohesion among its members. So under the rubric of "self-management," he neither designates someone to represent his authority on the team nor plays that role himself by actively participating in the team's deliberations.

Because the COO never enacts a clear authority boundary, the team gets lost in its own good feelings about itself. Its members end up creating too strong an identity boundary between themselves and the rest of the company. All of their energies go into maintaining that group identity, even at the price of suppressing the differences of skill and perspective they need to do the work. The team's strong identity boundary makes it impossible for members to establish the internal task and political boundaries necessary for functioning effectively. Put simply, the task force fails as a team because its members feel *too much* like a team.

Only when managers understand how boundaries interact in this way can they learn how to manage them. And executives must start by realizing that, like the COO, their most common mistakes are made at the authority boundary.

The Authority Vacuum

Senior executives know that in the new business environment the old authoritarian style—management by control—no longer works. Eager to encourage participation, teamwork, and employee empowerment, managers assume they must give up their own authority. But this decision has a paradoxical result. When managers abdicate authority, they cannot structure participation, teamwork, or empowerment effectively, which makes it impossible for their subordinates to be productive.

In the case of the unsuccessful product design task force, the vacuum of authority is filled by too much group cohesion. In other cases, teams respond to the authority vacuum by becoming passive. Instead of suppressing their differences and conflicts, team members are paralyzed by them. Because there is no strong authority to contain the inevitable tensions that increased participation necessarily generates, subordinates believe, quite rightly, that any conflicts will remain unresolved. Therefore, they dig in to protect their turf, and the conflicts become politicized. Since the senior executive is psychologically absent, people feel there is "no court of last resort," no guarantor to ensure that decisions will be fair.

Take, for example, what happened at a large multinational financial services company that was trying to design a new integrated information system. The company's executive group wanted to create a system that would link together the databases of the company's independent product lines and allow sales personnel to analyze a customer's financial needs and process applications immediately in the field. Such a system, they reasoned, would not only make for better and faster service, it would also position the company to take advantage of the next competitive frontier in their business—"cross-selling" a wide array of financial products to individual customers.

The executive group's strategic vision was sound, but their plan foundered on the messy details of implementation. From the very beginning, the ten-person committee responsible for developing a comprehensive technology plan became mired in conflicts. Some team members complained that the information systems division wasn't really committed to developing a new generation of code and programs. Product line representatives worried that they would lose the advantages of the highly customized information systems that each unit had developed over the years. And field service representatives were skeptical that headquarters could really accomplish such a complex and ambitious undertaking.

Frustrated by the committee's lack of progress and convinced that it "just didn't have the horsepower" to make the new system happen, the executive group set up a second, higher level team of three top managers—one from the information systems division and the other two from the company's major product groups. But this team was no more effective than the one it replaced. The IS representative complained that the product people didn't understand the complexities of the new technology. Meanwhile, the two product representatives accused the IS division of dragging its heels in order to protect its centralized control of the company's information systems. The executive group couldn't understand it. "Why aren't we getting any action?" they wondered.

What happened at this company illustrates what frequently occurs when a team of managers is asked to design new systems or products that promise to revolutionize core business relationships in a company. It's only natural that there will be conflicting perspectives and disagreements about critical issues. Only by confronting such conflicts can a team come up with a workable new approach.

A deeper analysis of the financial services company's conflicts revealed that beneath all the disagreement, the various groups represented on both committees shared a common feeling: they were anxious about the new strategic direction outlined by the company's senior executives. The company's success had traditionally been built on its organizational structure of highly autonomous product lines. Product managers functioned as independent entrepreneurs, free to extend and develop their lines without interference from headquarters. Naturally, they were afraid of losing this autonomy.

The problem was that no person on either the first or second committee had the authority to act as a tiebreaker when strong disagreements persisted. And neither team felt it could refer such deadlocks to the top management group. The chairperson of the first committee noted that he felt more like a "convener" than a team leader with decision-making power. And the smaller three-person team was explicitly designed as a group of equals with no single member in charge.

Why didn't the executive committee authorize someone to represent its new strategic intent on either committee? Upon reflection, senior managers realized they had shied away from exercising authority in the design process because they hadn't faced up to their *own* internal divisions about the risks of the company's new strategy. Specifically, the chief operating officer, who had strong ties to the independent product managers, was skeptical about the new direction. However, because he felt obliged to support the other two chief officers—another example of misplaced group cohesion—he remained silent. Of course, by failing to enact the authority boundary, this executive group created a vacuum that allowed political and identity differences to paralyze the work of both design committees.

As this example suggests, managers abdicate authority not just because they believe that's what flexible organizations require. On a deeper level, managers abdicate authority to defend themselves against their own anxieties. If conflict is an inherent feature of work in flexible organizations, so too is risk. In a business environment characterized by change and uncertainty, there is no guarantee that the decisions managers make or the strategic options they choose are the right ones. And as the competitive environment becomes more unforgiving, the consequences of failure become greater.

Often executives try to cope with this anxiety by focusing on tasks in a mechanistic way. They become enamored of elaborate methodologies for strategic planning. They systematically evaluate options and assess risks. These managers develop elegant plans for reorienting the strategic direction of the company. But while such detailed planning and analysis may help shape a decision, they cannot determine it. At some critical point, chief executives and their subordinates must move beyond the assessment of risks and make decisions in the face of considerable uncertainty. Indeed, senior management teams may be uncertain about *any* strategy's legitimacy and validity.

But in the end, a strategy's legitimacy rests on the personal authority of the chief executive—that is, on his or her ability to represent and embody the interests of the entire organization. If subordinates cannot identify psychologically with the chief executive as the representative of the whole, they will be unable to cede their autonomy to the CEO and, in effect, psychologically authorize him or her to lead. That means the best-laid plans and strategies will never be realized.

The real solution is for managers to exercise authority but in a new way. Authority in the corporation without boundaries is not about control but about *containment*—containment of the conflicts and anxieties that disrupt productive work. For instance, at the financial services company, the executive group took the simple step of appointing a new person as leader of the second design team. Putting an explicit representative of top management's authority on this second team was enough to keep the concerns over the company's new strategy from paralyzing everyone.

Managers need to be "present" for their subordinates in precisely this way, ready to resolve conflicts that cannot be managed by the group and to acknowledge honestly the anxiety inherent in working in an uncertain and risky environment. When they do so, managers can use their personal dealings with others to get people to do extraordinary things, even in the face of extremely difficult challenges.

Management as Containment: Downsizing with Dignity

For a portrait of a manager who works in this new way, consider how the vice president for human resources at a high-tech components manufacturer handled the downsizing of his own department. Once a leader in its business, this electronics company was facing a major crisis. Relentless innovation in the industry had led to falling prices and the growing technical obsolescence of the company's products, which produced a short-term squeeze on cash flow and a long-term threat to the company's survival.

What the company had to do to recapture market share was clear: dismantle its three independent business units, each with its own expensive staff functions, and replace them with a single multiproduct organization. The economies of scale made possible by eliminating duplicate support organizations would free up resources to invest in new R&D. Even more important, an integrated organization could respond to the growing trend toward technological integration in the company's three product lines as well as meet customer demands for integrated sales and service.

But shifting the strategic direction of the company necessarily came at a stiff price: a massive downsizing. The HR vice president faced the daunting task of managing the corporate reorganization and figuring out how his department could best serve the company's new business strategy, even as he laid off nearly 40% of his own staff, or about 20 people.

Downsizings bring the political boundary into play as perhaps no other management decision does. In the face of massive cuts, people struggle to defend their own interests, resources, and jobs. Concerned that the high level of retrenchment would demoralize his staff, the HR vice president was determined to manage the downsizing process in his own department in a way that would contain the politics of the situation.

The crucial decision he made was to ask subordinates to help him design a new and smaller human resources organization. The vice president believed that if people helped plan the cutback, the subsequent layoffs might feel less arbitrary and personal. Because they would understand the strategic logic behind the downsizing, those who left could do so with dignity, while those who remained would trust senior management's motives and plans.

At a meeting with his eight direct reports, the vice president divided them into two task forces. He asked both teams to come up with a wide range of possible configurations for the new human resources organization and to recruit some of their own subordinates as task force participants. The teams were to consider issues such as reporting relationships, spans of control, organizational structure, and new combinations of functions.

What's more, the vice president said, each of the proposed scenarios had to recognize four major constraints and challenges. First, head count had to drop by 40%. Yet the HR department also had to take on two new responsibilities: managing the companywide retrenchment plan and retraining the company's engineering personnel and sales force to function effectively in the new integrated organization. Finally, time was of the essence. The task forces had less than a month to come up with their proposals.

During this meeting, the HR vice president didn't just give orders. He did some extremely sophisticated work at both the authority and task boundaries. For example, his decision to ask the task forces for a variety of possible configurations rather than to come up with a single recommendation placed a clear authority boundary between himself and his subordinates. Because he took the responsibility for making the final decision, they didn't have to.

This authority boundary effectively contained the potentially destructive politics of the situation. Notice that the vice president didn't try to eliminate the politics altogether. Since each configuration was simply one among many possible alternatives, task force members felt free to advocate vigorously for a particular configuration without undermining the group's work. People could express their core interests without getting into a political deadlock. At the same time, the vice president also helped to contain politics by delineating a clear task boundary. By specifying a charge, a set of constraints, and a tight deadline, he helped task force members focus on the work.

The vice president also understood that he had to interact with his direct reports as individuals, not just as subordinates or technical specialists. The entire management team in the HR department confronted an unavoidable fact: some of the task force members were designing themselves out of a job. So at the end of the meeting, the vice president offered to meet with each of his subordinates privately. In the next two days, all eight met with the vice president for a confidential conversation about their own futures. What were their prospects? What would they do if the reorganization meant they had to leave the company? How might the vice president help them either to adapt to a new role at the company or to find a job elsewhere?

These conversations were painful but productive. They helped the subordinates feel that the vice president valued their thinking and would not discount their own personal dilemmas during the reorganization. The vice president showed that he could stay connected to them as people and to their own personal situations, even though he was the source of their immediate stress.

The private meetings also served another important purpose. By meeting with each of his staff members, the vice president helped to contain the uncertainty and risk, the difficulty and the pain, associated with downsizing. Naturally, his subordinates were preoccupied with their own personal welfare. But at the same time, they identified with the vice president's authority and wanted to satisfy him. Because he was connected to them emotionally and not only through a formal role relationship, they accepted him as their leader and were willing to do the work he expected of them. This emotional connection also helped them look beyond their immediate interests.

Over the next two weeks, the task forces produced a total of nine configurations. The discussions were stormy, and occasionally department heads strongly supported a particular design because it served their interests best. But their loyalty to the vice president and commitment to accomplishing the task he had given them meant they couldn't simply discount plans that might threaten their own jobs.

The vice president drew on a number of the plans to sketch his desired configuration. Then he worked with his direct reports, three of whom were to lose their jobs, to implement the new organizational design. While difficult for everyone involved, the implementation went smoothly. HR staff members believed in the new organizational structure. They were convinced that it fit well with the company's new business strategy and would enable the human resources department to do its job in an extremely difficult situation.

Even more important, people felt they had been treated with dignity. They had taken part in a "crisis team," doing important work at a critical time. They had helped the HR vice president and the company find the best solution to a difficult problem.

Getting Started: Feelings as Data

The HR vice president knew almost instinctively how to enact the right kind of boundary relationships with his subordinates. But what about the manager who is not so skilled? Where does he or she begin? To manage the new boundaries of flexible organizations, the best tools managers have are their own feelings.

At first glance, this claim may sound unlikely. After all, many managers tend to discount their feelings as having nothing to do with work. In particular, they view negative feelings as dangerous and disruptive. Either they ignore them, grit their teeth, and get on with the job, or they dismiss them as something merely personal, their own problem, unrelated to their work.

Yet anyone who has ever been part of a group that worked well together remembers how good that experience felt. When people have productive working relationships, they feel at ease, relaxed, and focused on their work. Work of this quality resembles a good conversation, in which people are "in sync" and everyone has something valuable to offer. When this happens, employees experience work as not only productive but also creative, innovative, and, quite simply, fun.

Similarly, when a work interaction has gone wrong and people are in the heat of a difficult situation, they often feel terrible. They become frustrated, angry, confused, and sometimes even ashamed. "I hate this," they find themselves thinking. "This isn't working." "I can't be productive." "I don't like the way 'X' is behaving." In such situations, people feel as if they are swimming against a strong current. Any sense of achievement or accomplishment they glean from work comes *despite* their participation in the organization or team, not because of it.

So too in the examples we have described. A common denominator in all of them is the presence of strong feelings: the engineer's defensiveness when questioned by a shop-floor worker; the shame and anger of that worker, feeling put down by the engineer; the contempt that product design team members felt toward anyone not part of their intense team experience; the anxieties that led to the paralyzing conflicts at the financial services company.

Such feelings aren't just the inevitable emotional residue of human work relationships. They are *data*, valuable clues to the dynamics of boundary relationships. In this respect, feelings are an aid to thinking and to managing; they are a real part of real work. Like the human resources vice president, the best managers understand this intuitively. They not only manage with their heads but also with their gut feelings.

To be good boundary managers, executives must be able to decipher frustrating and difficult personal relationships and diagnose why they have gone wrong. Doing so requires acknowledging their own often intense personal responses to work situations. Much as managers learn how to make subtle distinctions when interpreting empirical data, boundary management also requires developing a more precise language for describing the feelings people experience at work.

But at the same time, managers also need to know how to distance themselves from their own experience and feelings—in a sense, to depersonalize them—in order to see how their own responses are symptoms of a broader group process. In fact, the stronger the negative feelings people have about a work interaction, the less likely those feelings are "just personal" and the more likely they are a symptom of a real organizational problem. Feelings are important signals to managers that they must step back and examine their work relationships.

For example, had the engineer been able to ask himself, "Why do I feel so attacked?," then his encounter with the worker might have had a more productive outcome. Similarly, the defensiveness of the product design team in response to questions from top management should have been a sign to team members and the COO that there was a fundamental structural problem in the team's relationship to the executive group. Finally, had senior executives at the financial services company understood their frustration with the first design committee as a symptom of a failed work process, rather than as evidence that the individuals on the team lacked "horsepower," the managers might have avoided making the same mistake twice.

Using one's own feelings to diagnose relationships on the job is hard work. Yet an awareness of feelings, one's own and those of others, is crucial to making flexible organizations work. It is the way to discover the boundaries people need in relationships to achieve their best. It is at the very heart of management in the "corporation without boundaries."

2
Teaching Smart People How to Learn

Chris Argyris

Any company that aspires to succeed in the tougher business environment of the 1990s must first resolve a basic dilemma: success in the marketplace increasingly depends on learning, yet most people don't know how to learn. What's more, those members of the organization that many assume to be the best at learning are, in fact, not very good at it. I am talking about the well-educated, high-powered, high-commitment professionals who occupy key leadership positions in the modern corporation.

Most companies not only have tremendous difficulty addressing this learning dilemma; they aren't even aware that it exists. The reason: they misunderstand what learning is and how to bring it about. As a result, they tend to make two mistakes in their efforts to become a learning organization.

First, most people define learning too narrowly as mere "problem solving," so they focus on identifying and correcting errors in the external environment. Solving problems is important. But if learning is to persist, managers and employees must also look inward. They need to reflect critically on their own behavior, identify the ways they often inadvertently contribute to the organization's problems, and then change how they act. In particular, they must learn how the very way they go about defining and solving problems can be a source of problems in its own right.

I have coined the terms "single loop" and "double loop" learning to capture this crucial distinction. To give a simple analogy: a thermostat that automatically turns on the heat whenever the temperature in a room drops below 68 degrees is a good example of single-loop learning. A thermostat that could ask, "Why am I set at 68 degrees?" and then explore whether or not some other temperature might more economically achieve the goal of heating the room would be engaging in double-loop learning.

Highly skilled professionals are frequently very good at single-loop learning. After all, they have spent much of their lives acquiring academic credentials, mastering one or a number of intellectual disciplines, and applying those disciplines to solve real-world problems. But ironically, this very fact helps explain why professionals are often so bad at double-loop learning.

Put simply, because many professionals are almost always successful at what they do, they rarely experience failure. And because they have rarely failed, they have never learned how to learn from failure. So whenever their single-loop learning strategies go wrong, they become defensive, screen out criticism, and put the "blame" on anyone and everyone but themselves. In short, their ability to learn shuts down precisely at the moment they need it the most.

The propensity among professionals to behave defensively helps shed light on the second mistake that companies make about learning. The common assumption is that getting people to learn is largely a matter of motivation. When people have the right attitudes and commitment, learning automatically follows. So companies focus on creating new organizational structures—compensation programs, performance reviews, corporate cultures, and the like—that are designed to create motivated and committed employees.

But effective double-loop learning is not simply a function of how people feel. It is a reflection of how they think—that is, the cognitive rules or reasoning they use to design and implement their actions. Think of these rules as a kind of "master program" stored in the brain, governing all behavior. Defensive reasoning can block learning even when the individual commitment to it is high, just as a computer program with hidden bugs can produce results exactly the opposite of what its designers had planned.

Companies can learn how to resolve the learning dilemma. What it takes is to make the ways managers and employees reason about their behavior a focus of organizational learning and continuous improvement programs. Teaching people how to reason about their behavior in new and more effective ways breaks down the defenses that block learning.

All of the examples that follow involve a particular kind of professional: fast-track consultants at major management consulting companies. But the implications of my argument go far beyond this specific occupational group. The fact is, more and more jobs—no matter what the title—are taking on the contours of "knowledge work." People at all levels of the organization must combine the mastery of some highly specialized technical expertise with the ability to work effectively in teams, form productive relationships with clients and customers, and critically reflect on and then change their own organizational practices. And the nuts and bolts of management—whether of high-powered consultants or service representatives, senior managers or factory technicians—increasingly consists of guiding and integrating the autonomous but interconnected work of highly skilled people.

How Professionals Avoid Learning

For 15 years, I have been conducting in-depth studies of management consultants. I decided to study consultants for a few simple reasons. First, they are the epitome of the highly educated professionals who play an increasingly central role in all organizations. Almost all of the consultants I've studied have MBAs from the top three or four U.S. business schools. They are also highly committed to their work. For instance, at one company, more than 90% of the consultants responded in a survey that they were "highly satisfied" with their jobs and with the company.

I also assumed that such professional consultants would be good at learning. After all, the essence of their job is to teach others how to do things differently. I found, however, that these consultants embodied the learning dilemma. The most enthusiastic about continuous improvement in their own organizations, they were also often the biggest obstacle to its complete success.

As long as efforts at learning and change focused on external organizational factors—job redesign, compensation programs, performance reviews, and leadership training—the professionals were enthusiastic participants. Indeed, creating new systems and structures was precisely the kind of challenge that well-educated, highly motivated professionals thrived on.

And yet the moment the quest for continuous improvement turned to the professionals' *own* performance, something went wrong. It wasn't a matter of bad attitude. The professionals' commitment to excellence was genuine, and the vision of the company was clear. Nevertheless, continuous improvement did not persist. And the longer the continuous improvement efforts continued, the greater the likelihood that they would produce ever-diminishing returns.

What happened? The professionals began to feel embarrassed. They were threatened by the prospect of critically examining their own role in the organization. Indeed, because they were so well paid (and generally believed that their employers were supportive and fair), the idea that their performance might not be at its best made them feel guilty.

Far from being a catalyst for real change, such feelings caused most to react defensively. They projected the blame for any problems away from themselves and onto what they said were unclear goals, insensitive and unfair leaders, and stupid clients.

Consider this example. At a premier management consulting company, the manager of a case team called a meeting to examine the team's performance on a recent consulting project. The client was largely satisfied and had given the team relatively high marks, but the manager believed the team had not created the value added that it was capable of and that the consulting company had promised. In the spirit of continuous improvement, he felt that the team could do better. Indeed, so did some of the team members.

The manager knew how difficult it was for people to reflect critically on their own work performance, especially in the presence of their manager, so he took a number of steps to make possible a frank and open discussion. He invited to the meeting an outside consultant whom team members knew and trusted—"just to keep me honest," he said. He also agreed to have the entire meeting tape-recorded. That way, any subsequent confusions or disagreements about what went on at the meeting could be checked against the transcript. Finally, the manager opened the meeting by emphasizing that no subject was off limits—including his own behavior.

"I realize that you may believe you cannot confront me," the manager said. "But I encourage you to challenge me. You have a responsibility to tell me where you think the leadership made mistakes, just as I have the responsibility to identify any I believe you made. And all of us must acknowledge our own mistakes. If we do not have an open dialogue, we will not learn."

The professionals took the manager up on the first half of his invitation but quietly ignored the second. When asked to pinpoint the key problems in the experience with the client, they looked entirely outside themselves. The clients were uncooperative and arrogant. "They didn't think we could help them." The team's own managers were unavailable and poorly prepared. "At times, our managers were not up to speed before they walked into the client meetings." In effect, the professionals asserted that they were helpless to act differently—not because of any limitations of their own but because of the limitations of others.

The manager listened carefully to the team members and tried to respond to their criticisms. He talked about the mistakes that he had made during the consulting process. For example, one professional objected to the way the manager had run the project meetings. "I see that the way I asked questions closed down discussions," responded the manager. "I didn't mean to do that, but I can see how you might have believed that I had already made up my mind." Another team member complained that the manager had caved in to pressure from his superior to produce the project report far too quickly, considering the team's heavy work load. "I think that it was my responsibility to have said no," admitted the manager. "It was clear that we all had an immense amount of work."

Finally, after some three hours of discussion about his own behavior, the manager began to ask the team members if there were any errors *they* might have made. "After all," he said, "this client was not different from many others. How can we be more effective in the future?"

The professionals repeated that it was really the clients' and their own managers' fault. As one put it, "They have to be open to change and want to learn." The more the manager tried to get the team to examine its own responsibility for the outcome, the more the professionals bypassed his concerns. The best one team member could suggest was for the case team to "promise less"—implying that there was really no way for the group to improve its performance.

The case team members were reacting defensively to protect themselves, even though their manager was not acting in ways that an outsider would consider threatening. Even if there were some truth to their charges—the clients may well have been arrogant and closed, their own managers distant—the *way* they presented these claims was guaranteed to stop learning. With few exceptions, the professionals made attributions about the behavior of the clients and the managers but never publicly tested their claims. For instance, they said that the clients weren't motivated to learn but never really presented any evidence supporting that assertion. When their lack of concrete evidence was pointed out to them, they simply repeated their criticisms more vehemently.

If the professionals had felt so strongly about these issues, why had they never mentioned them during the project? According to the professionals, even this was the fault of others. "We didn't want to alienate the client," argued one. "We didn't want to be seen as whining," said another.

The professionals were using their criticisms of others to protect themselves from the potential embarrassment of having to admit that perhaps they too had contributed to the team's less-than-perfect performance. What's more, the fact that they kept repeating their defensive actions in the face of the manager's efforts to turn the group's attention to its own role shows that this defensiveness had become a reflexive routine. From the professionals' perspective, they weren't resisting; they were focusing on the "real" causes. Indeed, they were to be respected, if not congratulated, for working as well as they did under such difficult conditions.

The end result was an unproductive parallel conversation. Both the manager and the professionals were candid; they expressed their views forcefully. But they talked past each other, never finding a common language to describe what had happened with the client. The professionals kept insisting that the fault lay with others. The manager kept trying, unsuccessfully, to get the professionals to see how they contributed to the state of affairs they were criticizing. The dialogue of this parallel conversation looks like this:

PROFESSIONALS: "The clients have to be open. They must want to change."

MANAGER: "It's our task to help them see that change is in their interest."

PROFESSIONALS: "But the clients didn't agree with our analyses."

MANAGER: "If they didn't think our ideas were right, how might we have convinced them?"

PROFESSIONALS: "Maybe we need to have more meetings with the client."

MANAGER: "If we aren't adequately prepared and if the clients don't think we're credible, how will more meetings help?"

PROFESSIONALS: "There should be better communication between case team members and management."

MANAGER: "I agree. But professionals should take the initiative to educate the manager about the problems they are experiencing."

PROFESSIONALS: "Our leaders are unavailable and distant."

MANAGER: "How do you expect us to know that if you don't tell us?"

Conversations such as this one dramatically illustrate the learning dilemma. The problem with the professionals' claims is not that they are wrong but that they aren't useful. By constantly turning the focus away from their own behavior to that of others, the professionals bring learning to a grinding halt. The manager understands the trap but does not know how to get out of it. To learn how to do that requires going deeper into the dynamics of defensive reasoning—and into the special causes that make professionals so prone to it.

Defensive Reasoning and the Doom Loop

What explains the professionals' defensiveness? Not their attitudes about change or commitment to continuous improvement; they really wanted to work more effectively. Rather, the key factor is the way they reasoned about their behavior and that of others.

It is impossible to reason anew in every situation. If we had to think through all the possible responses every time someone asked, "How are you?" the world would pass us by. Therefore, everyone develops a theory of action—a set of rules that individuals use to design and implement their own behavior as well as to understand the behavior of others. Usually, these theories of actions become so taken for granted that people don't even realize they are using them.

One of the paradoxes of human behavior, however, is that the master program people actually use is rarely the one they think they use. Ask people in an interview or questionnaire to articulate the rules they use to govern their actions, and they will give you what I call their "espoused" theory of action. But observe these same people's behavior, and you will quickly see that this espoused theory has very little to do with how they actually behave. For example, the professionals on the case team said they believed in continuous improvement, and yet they consistently acted in ways that made improvement impossible.

When you observe people's behavior and try to come up with rules that would make sense of it, you discover a very different theory of action—what I call the individual's "theory-in-use." Put simply, people consistently act inconsistently, unaware of the contradiction between their espoused theory and their theory-in-use, between the way they think they are acting and the way they really act.

What's more, most theories-in-use rest on the same set of governing values. There seems to be a universal human tendency to design one's actions consistently according to four basic values:

1. To remain in unilateral control;
2. To maximize "winning" and minimize "losing";
3. To suppress negative feelings; and
4. To be as "rational" as possible—by which people mean defining clear objectives and evaluating their behavior in terms of whether or not they have achieved them.

The purpose of all these values is to avoid embarrassment or threat, feeling vulnerable or incompetent. In this respect, the master program that most people use is profoundly defensive. Defensive reasoning encourages individuals to keep private the premises, inferences, and conclusions that shape their behavior and to avoid testing them in a truly independent, objective fashion.

Because the attributions that go into defensive reasoning are never really tested, it is a closed loop, remarkably impervious to conflicting points of view. The inevitable response to the observation that somebody is reasoning defensively is yet more defensive reasoning. With the case team, for example, whenever anyone pointed out the professionals' defensive behavior to them, their initial reaction was to look for the cause in somebody else—clients who were so sensitive that they would have been alienated if the consultants had criticized them or a manager so weak that he couldn't have taken it had the consultants raised their concerns with him. In other words, the case team members once again denied their own responsibility by externalizing the problem and putting it on someone else.

In such situations, the simple act of encouraging more open inquiry is often attacked by others as "intimidating." Those who do the attacking deal with their feelings about possibly being wrong by blaming the more open individual for arousing these feelings and upsetting them.

Needless to say, such a master program inevitably short-circuits learning. And for a number of reasons unique to their psychology, well-educated professionals are especially susceptible to this.

Nearly all the consultants I have studied have stellar academic records. Ironically, their very success at education helps explain the problems they have with learning. Before they enter the world of work, their lives are primarily full of successes, so they have rarely experienced the embarrassment and sense of threat that comes with failure. As a result, their defensive reasoning has rarely been activated. People who rarely experience failure, however, end up not knowing how to deal with it effectively. And this serves to reinforce the normal human tendency to reason defensively.

In a survey of several hundred young consultants at the organizations I have been studying, these professionals describe themselves as driven internally by an unrealistically high ideal of performance: "Pressure on the job is self-imposed." "I must not only do a good job; I must also be the best." "People around here are very bright and hardworking; they are highly motivated to do an outstanding job." "Most of us want not only to succeed but also to do so at maximum speed."

These consultants are always comparing themselves with the best around them and constantly trying to better their own performance. And yet they do not appreciate being required to compete openly with each other. They feel it is somehow inhumane. They prefer to be the individual contributor—what might be termed a "productive loner."

Behind this high aspiration for success is an equally high fear of failure and a propensity to feel shame and guilt when they do fail to meet their high standards. "You must avoid mistakes," said one. "I hate making them. Many of us fear failure, whether we admit it or not."

To the extent that these consultants have experienced success in their lives, they have not had to be concerned about failure and the attendant feelings of shame and guilt. But to exactly the same extent, they also have never developed the tolerance for feelings of failure or the skills to deal with these feelings. This in turn has led them not only to fear failure but also to fear the fear of failure itself. For they know that they will not cope with it superlatively—their usual level of aspiration.

The consultants use two intriguing metaphors to describe this phenomenon. They talk about the "doom loop" and "doom zoom." Often, consultants will perform well on the case team, but because they don't do the jobs perfectly or receive accolades from their managers, they go into a doom loop of despair. And they don't ease into the doom loop, they zoom into it.

As a result, many professionals have extremely "brittle" personalities. When suddenly faced with a situation they cannot immediately handle, they tend to fall apart. They cover up their distress in front of the client. They talk about it constantly with their fellow case team members. Interestingly, these conversations commonly take the form of bad-mouthing clients.

Such brittleness leads to an inappropriately high sense of despondency or even despair when people don't achieve the high levels of performance they aspire to. Such despondency is rarely psychologically devastating, but when combined with defensive reasoning, it can result in a formidable predisposition against learning.

There is no better example of how this brittleness can disrupt an organization than performance evaluations. Because it represents the one moment when a professional must measure his or her own behavior against some formal standard, a performance evaluation is almost tailor-made to push a professional into the doom loop. Indeed, a poor evaluation can reverberate far beyond the particular individual involved to spark defensive reasoning throughout an entire organization.

At one consulting company, management established a new performance-evaluation process that was designed to make evaluations both more objective and more useful to those being evaluated. The consultants participated in the design of the new system and in general were enthusiastic because it corresponded to their espoused values of objectivity and fairness. A brief two years into the new process, however, it had become the object of dissatisfaction. The catalyst for this about-face was the first unsatisfactory rating.

Senior managers had identified six consultants whose performance they considered below standard. In keeping with the new evaluation process, they did all they could to communicate their concerns to the six and to help them improve. Managers met with each individual separately for as long and as often as the professional requested to explain the reasons behind the rating and to discuss what needed to be done to improve—but to no avail. Performance continued at the same low level and, eventually, the six were let go.

When word of the dismissal spread through the company, people responded with confusion and anxiety. After about a dozen consultants angrily complained to management, the CEO held two lengthy meetings where employees could air their concerns.

At the meetings, the professionals made a variety of claims. Some said the performance-evaluation process was unfair because judgments were subjective and biased and the criteria for minimum performance unclear. Others suspected that the real cause for the dismissals was economic and that the performance-evaluation procedure was just a fig leaf to hide the fact that the company was in trouble. Still others argued that the evaluation process was antilearning. If the company were truly a learning organization, as it claimed, then people performing below the minimum standard should be taught how to reach it. As one professional put it: "We were told that the company did not have an up-or-out policy. Up-or-out is inconsistent with learning. You misled us."

The CEO tried to explain the logic behind management's decision by grounding it in the facts of the case and by asking the professionals for any evidence that might contradict these facts.

Is there subjectivity and bias in the evaluation process? Yes, responded the CEO, but "we strive hard to reduce them. We are constantly trying to improve the process. If you have any ideas, please tell us. If you know of someone treated unfairly, please bring it up. If any of you feel that you have been treated unfairly, let's discuss it now or, if you wish, privately."

Is the level of minimum competence too vague? "We are working to define minimum competence more clearly," he answered. "In the case of the six, however, their performance was so poor that it wasn't difficult to reach a decision." Most of the six had received timely feedback about their problems. And in the two cases where people had not, the reason was that they had never taken the responsibility to seek out evaluations—and, indeed, had actively avoided them. "If you have any data to the contrary," the CEO added, "let's talk about it."

Were the six asked to leave for economic reasons? No, said the CEO. "We have more work than we can do, and letting professionals go is extremely costly for us. Do any of you have any information to the contrary?"

As to the company being antilearning, in fact, the entire evaluation process was designed to encourage learning. When a professional is performing below the minimum level, the CEO explained, "we jointly design remedial experiences with the individual. Then we look for signs of improvement. In these cases, either the professionals were reluctant to take on such assignments or they repeatedly failed when they did. Again, if you have information or evidence to the contrary, I'd like to hear about it."

The CEO concluded: "It's regrettable, but sometimes we make mistakes and hire the wrong people. If individuals don't produce and repeatedly prove themselves unable to improve, we don't know what else to do except dismiss them. It's just not fair to keep poorly per-forming individuals in the company. They earn an unfair share of the financial rewards."

Instead of responding with data of their own, the professionals simply repeated their accusations but in ways that consistently contra-dicted their claims. They said that a genuinely fair evaluation process would contain clear and documentable data about performance—but they were unable to provide firsthand examples of the unfairness that they implied colored the evaluation of the six dismissed employees. They argued that people shouldn't be judged by inferences uncon-nected to their actual performance—but they judged management in precisely this way. They insisted that management define clear, objec-tive, and unambiguous performance standards—but they argued that any humane system would take into account that the performance of a professional cannot be precisely measured. Finally, they presented themselves as champions of learning—but they never proposed any criteria for assessing whether an individual might be unable to learn.

In short, the professionals seemed to hold management to a differ-ent level of performance than they held themselves. In their conver-sation at the meetings, they used many of the features of ineffective evaluation that they condemned—the absence of concrete data, for example, and the dependence on a circular logic of "heads we win, tails you lose." It is as if they were saying, "Here are the features of a fair performance-evaluation system. You should abide by them. But we don't have to when we are evaluating you."

Indeed, if we were to explain the professionals' behavior by articu-lating rules that would have to be in their heads in order for them to act the way they did, the rules would look something like this:

1. When criticizing the company, state your criticism in ways that you believe are valid—but also in ways that prevent others from deciding for themselves whether your claim to validity is correct.

2. When asked to illustrate your criticisms, don't include any data that others could use to decide for themselves whether the illustrations are valid.

3. State your conclusions in ways that disguise their logical implications. If others point out those implications to you, deny them.

Of course, when such rules were described to the professionals, they found them abhorrent. It was inconceivable that these rules might explain their actions. And yet in defending themselves against this observation, they almost always inadvertently confirmed the rules.

Learning How to Reason Productively

If defensive reasoning is as widespread as I believe, then focusing on an individual's attitudes or commitment is never enough to produce real change. And as the previous example illustrates, neither is creating new organizational structures or systems. The problem is that even when people are genuinely committed to improving their performance and management has changed its structures in order to encourage the "right" kind of behavior, people still remain locked in defensive reasoning. Either they remain unaware of this fact, or if they do become aware of it, they blame others.

There is, however, reason to believe that organizations can break out of this vicious circle. Despite the strength of defensive reasoning, people genuinely strive to produce what they intend. They value acting competently. Their self-esteem is intimately tied up with behaving consistently and performing effectively. Companies can use these universal human tendencies to teach people how to reason in a new way—in effect, to change the master programs in their heads and thus reshape their behavior.

People can be taught how to recognize the reasoning they use when they design and implement their actions. They can begin to identify the inconsistencies between their espoused and actual theories of action. They can face up to the fact that they unconsciously design and implement actions that they do not intend. Finally, people can learn how to identify what individuals and groups do to create organizational defenses and how these defenses contribute to an organization's problems.

Once companies embark on this learning process, they will discover that the kind of reasoning necessary to reduce and overcome organizational defenses is the same kind of "tough reasoning" that underlies the effective use of ideas in strategy, finance, marketing, manufacturing, and other management disciplines. Any sophisticated strategic analysis, for example, depends on collecting valid data, analyzing it carefully, and constantly testing the inferences drawn from the data. The toughest tests are reserved for the conclusions. Good strategists make sure that their conclusions can withstand all kinds of critical questioning.

So too with productive reasoning about human behavior. The standard of analysis is just as high. Human resources programs no longer need to be based on "soft" reasoning but should be as analytical and as data-driven as any other management discipline.

Of course, that is not the kind of reasoning the consultants used when they encountered problems that were embarrassing or threatening. The data they collected was hardly objective. The inferences they made rarely became explicit. The conclusions they reached were largely self-serving, impossible for others to test, and as a result, "self-sealing," impervious to change.

How can an organization begin to turn this situation around, to teach its members how to reason productively? The first step is for managers at the top to examine critically and change their own theories-in-use. Until senior managers become aware of how they reason defensively and the counterproductive consequences that result, there will be little real progress. Any change activity is likely to be just a fad.

Change has to start at the top because otherwise defensive senior managers are likely to disown any transformation in reasoning patterns coming from below. If professionals or middle managers begin to change the way they reason and act, such changes are likely to appear strange—if not actually dangerous—to those at the top. The result is an unstable situation where senior managers still believe that it is a sign of caring and sensitivity to bypass and cover up difficult issues, while their subordinates see the very same actions as defensive.

The key to any educational experience designed to teach senior managers how to reason productively is to connect the program to real business problems. The best demonstration of the usefulness of productive reasoning is for busy managers to see how it can make a direct difference in their own performance and in that of the organization. This will not happen overnight. Managers need plenty of opportunity to practice the new skills. But once they grasp the powerful impact that productive reasoning can have on actual performance, they will have a strong incentive to reason productively not just in a training session but in all their work relationships.

One simple approach I have used to get this process started is to have participants produce a kind of rudimentary case study. The subject is a real business problem that the manager either wants to deal with or has tried unsuccessfully to address in the past. Writing the actual case usually takes less than an hour. But then the case becomes the focal point of an extended analysis.

For example, a CEO at a large organizational-development consulting company was preoccupied with the problems caused by the intense competition among the various business functions represented by his four direct reports. Not only was he tired of having the problems dumped in his lap, but he was also worried about the impact the interfunctional conflicts were having on the organization's flexibility.

He had even calculated that the money being spent to iron out disagreements amounted to hundreds of thousands of dollars every year. And the more fights there were, the more defensive people became, which only increased the costs to the organization.

In a paragraph or so, the CEO described a meeting he intended to have with his direct reports to address the problem. Next, he divided the paper in half, and on the right-hand side of the page, he wrote a scenario for the meeting—much like the script for a movie or play—describing what he would say and how his subordinates would likely respond. On the left-hand side of the page, he wrote down any thoughts and feelings that he would be likely to have during the meeting but that he wouldn't express for fear they would derail the discussion.

But instead of holding the meeting, the CEO analyzed this scenario *with* his direct reports. The case became the catalyst for a discussion in which the CEO learned several things about the way he acted with his management team.

He discovered that his four direct reports often perceived his conversations as counterproductive. In the guise of being "diplomatic," he would pretend that a consensus about the problem existed, when in fact none existed. The unintended result: instead of feeling reassured, his subordinates felt wary and tried to figure out "what is he *really* getting at."

The CEO also realized that the way he dealt with the competitiveness among department heads was completely contradictory. On the one hand, he kept urging them to "think of the organization as a whole." On the other, he kept calling for actions—department budget cuts, for example—that placed them directly in competition with each other.

Finally, the CEO discovered that many of the tacit evaluations and attributions he had listed turned out to be wrong. Since he had never expressed these assumptions, he had never found out just how wrong they were. What's more, he learned that much of what he thought he was hiding came through to his subordinates anyway—but with the added message that the boss was covering up.

The CEO's colleagues also learned about their own ineffective behavior. They learned by examining their own behavior as they tried to help the CEO analyze his case. They also learned by writing and analyzing cases of their own. They began to see that they too tended to bypass and cover up the real issues and that the CEO was often aware of it but did not say so. They too made inaccurate attributions and evaluations that they did not express. Moreover, the belief that they had to hide important ideas and feelings from the CEO and from each other in order not to upset anyone turned out to be mistaken. In the context of the case discussions, the entire senior management team was quite willing to discuss what had always been undiscussable.

In effect, the case study exercise legitimizes talking about issues that people have never been able to address before. Such a discussion can be emotional—even painful. But for managers with the courage to persist, the payoff is great: management teams and entire organizations work more openly and more effectively and have greater options for behaving flexibly and adapting to particular situations.

When senior managers are trained in new reasoning skills, they can have a big impact on the performance of the entire organization— even when other employees are still reasoning defensively. The CEO who led the meetings on the performance-evaluation procedure was able to defuse dissatisfaction because he didn't respond to professionals' criticisms in kind but instead gave a clear presentation of relevant data. Indeed, most participants took the CEO's behavior to be a sign that the company really acted on the values of participation and employee involvement that it espoused.

Of course, the ideal is for all the members of an organization to learn how to reason productively. This has happened at the company where the case team meeting took place. Consultants and their managers are now able to confront some of the most difficult issues of the consultant-client relationship. To get a sense of the difference productive reasoning can make, imagine how the original conversation between the manager and case team might have gone had everyone engaged in effective reasoning. (The following dialogue is based on actual sessions I have attended with other case teams at the same company since the training has been completed.)

First, the consultants would have demonstrated their commitment to continuous improvement by being willing to examine their own role in the difficulties that arose during the consulting project. No doubt they would have identified their managers and the clients as part of the problem, but they would have gone on to admit that they had contributed to it as well. More important, they would have agreed with the manager that as they explored the various roles of clients, managers, and professionals, they would make sure to test any evaluations or attributions they might make against the data. Each individual would have encouraged the others to question his or her reasoning. Indeed, they would have insisted on it. And in turn, everyone would have understood that act of questioning not as a sign of mistrust or an invasion of privacy but as a valuable opportunity for learning.

The conversation about the manager's unwillingness to say no might look something like this:

PROFESSIONAL #1: "One of the biggest problems I had with the way you managed this case was that you seemed to be unable to say no when either the client or your superior made unfair demands." [Gives an example.]

PROFESSIONAL #2: "I have another example to add. [Describes a second example.] But I'd also like to say that we never really told you how we felt about this. Behind your back we were bad-mouthing you—you know, 'he's being such a wimp'—but we never came right out and said it."

MANAGER: "It certainly would have been helpful if you had said something. Was there anything I said or did that gave you the idea that you had better not raise this with me?"

PROFESSIONAL #3: "Not really. I think we didn't want to sound like we were whining."

MANAGER: "Well, I certainly don't think you sound like you're whining. But two thoughts come to mind. If I understand you correctly, you *were* complaining, but the complaining about me and my inability to say no was covered up. Second, if we had discussed this, I might have gotten the data I needed to be able to say no."

Notice that when the second professional describes how the consultants had covered up their complaints, the manager doesn't criticize her. Rather, he rewards her for being open by responding in kind. He focuses on the ways that he too may have contributed to the cover-up. Reflecting undefensively about his own role in the problem then makes it possible for the professionals to talk about their fears of appearing to be whining. The manager then agrees with the professionals that they shouldn't become complainers. At the same time, he points out the counterproductive consequences of covering up their complaints.

Another unresolved issue in the case team meeting concerned the supposed arrogance of the clients. A more productive conversation about that problem might go like this:

MANAGER: "You said that the clients were arrogant and uncooperative. What did they say and do?"

PROFESSIONAL #1: "One asked me if I had ever met a payroll. Another asked how long I've been out of school."

PROFESSIONAL #2: "One even asked me how old I was!"

PROFESSIONAL #3: "That's nothing. The worst is when they say that all we do is interview people, write a report based on what they tell us, and then collect our fees."

MANAGER: "The fact that we tend to be so young is a real problem for many of our clients. They get very defensive about it. But I'd like to explore whether there is a way for them to freely express their views without our getting defensive.

"What troubled me about your original responses was that you assumed you were right in calling the clients stupid. One thing I've noticed about consultants—in this company and others—is that we tend to defend ourselves by bad-mouthing the client."

PROFESSIONAL #1: "Right. After all, if they are genuinely stupid, then it's obviously not our fault that they aren't getting it!"

PROFESSIONAL #2: "Of course, that stance is antilearning and overprotective. By assuming that they can't learn, we absolve ourselves from having to."

PROFESSIONAL #3: "And the more we all go along with the bad-mouthing, the more we reinforce each other's defensiveness."

MANAGER: "So what's the alternative? How can we encourage our clients to express their defensiveness and at the same time constructively build on it?"

PROFESSIONAL #1: "We all know that the real issue isn't our age; it's whether or not we are able to add value to the client's organization. They should judge us by what we produce. And if we aren't adding value, they should get rid of us—no matter how young or old we happen to be."

MANAGER: "Perhaps that is exactly what we should tell them."

In both these examples, the consultants and their manager are doing real work. They are learning about their own group dynamics and addressing some generic problems in client-consultant relationships. The insights they gain will allow them to act more effectively in the future—both as individuals and as a team. They are not just solving problems but developing a far deeper and more textured understanding of their role as members of the organization. They are laying the groundwork for continuous improvement that is truly continuous. They are learning how to learn.

3
Values Make the Company: An Interview with Robert Haas

Robert Howard

As chairman and CEO of Levi Strauss & Co., Robert D. Haas has inherited a dual legacy. Ever since its founding in 1850, the San Francisco-based apparel manufacturer has been famous for combining strong commercial success with a commitment to social values and to its work force.

Achieving both goals was relatively easy throughout much of the postwar era, when the company's main product—Levi's jeans—became an icon of American pop culture and sales surged on the demographic wave of the expanding baby boom. But in the uncertain economic climate of the 1980s, Haas and his management team have had to rethink every facet of the business—including its underlying values.

Since his appointment as CEO in 1984, Haas has redefined the company's business strategy; created a flatter organization, including the painful step of cutting the work force by one-third; and invested heavily in new product development, marketing, and technology. In 1985, he and his team took the company private in one of the most successful management-led LBOs of the 1980s. And in 1987, he oversaw the development of the Levi Strauss Aspirations Statement, a major initiative to define the shared values that will guide both management and the work force.

Many CEOs talk about values, but few have gone to the lengths Haas has to bring them to the very center of how he runs the business. The Aspirations Statement is shaping how the company defines occupational roles and responsibilities, conducts performance evaluations, trains new employees, organizes work, and makes business decisions.

The result is a remarkably flexible and innovative company, despite its age and size. Levi is a pioneer in using electronic networks to link the company more closely to its suppliers and retailers. The Dockers line of clothing, introduced in 1986, has been one of the fastest-growing new products in apparel industry history. And the company has made a successful major push into global markets. In 1989, international operations accounted for 34% of Levi's total sales and 45% of pretax operating profit.

Levi's financial results have also been extraordinary. From 1985 to 1989, sales increased 31% to $3.6 billion. And profits have risen fivefold to $272 million.

Meanwhile, the company has stayed true to its traditional commitment to social issues even as it has updated that commitment to reflect the economic and social realities of a new era. Levi Strauss has an exemplary record on issues ranging from work force diversity to benefits for workers dislocated by plant closings and technological change. Haas himself is the foremost corporate spokesperson on the responsibilities of business in the AIDS crisis.

Reinventing the Levi Strauss heritage has a special meaning for Haas. He is the great-great-grandnephew of the company founder, and his uncle, father, and grandfather all led the company before him. He joined Levi Strauss in 1973 and has served in a variety of leadership positions, including senior vice president of corporate planning and policy, president of the operating groups, and executive vice president and chief operating officer. He has also worked as an associate at McKinsey & Co. and spent two years as a Peace Corps volunteer in the Ivory Coast.

The interview was conducted by HBR associate editor Robert Howard.

HBR: Levi Strauss has long had a reputation for its social responsibility. Why are you placing so much emphasis on defining the company's values now?

Robert Haas: Levi has always treated people fairly and cared about their welfare. The usual term is "paternalism." But it is more than paternalism, really—a genuine concern for people and a recognition that people make this business successful.

In the past, however, that tradition was viewed as something separate from how we ran the business. We always talked about the "hard stuff" and the "soft stuff." The soft stuff was the company's commitment to our work force. And the hard stuff was what really mattered: getting pants out the door.

What we've learned is that the soft stuff and the hard stuff are becoming increasingly intertwined. A company's values—what it stands for, what its people believe in—are crucial to its competitive success. Indeed, values drive the business.

What is happening in your environment to bring you to that conclusion?

Traditionally, the business world had clear boundaries. Geographical or regional borders defined the marketplace. Distinctions between suppliers and customers, workers and managers, were well defined. Once you had a strong market position, you could go on for a long time just on inertia. You could have a traditional, hierarchical, command-and-control organization, because change happened so slowly.

People's expectations for work were also narrowly defined. They gave their loyalty and their efforts in exchange for being taken care of. They expected information and commands to come down from on high, and they did what they were told.

As a result of all the tumult of the 1980s—increased competition, corporate restructurings, the globalization of enterprises, a new generation entering the work force—those traditional boundaries and expectations are breaking down.

What do those changes mean for leadership?

There is an enormous diffusion of power. If companies are going to react quickly to changes in the marketplace, they have to put more and more accountability, authority, and information into the hands of the people who are closest to the products and the customers. That requires new business strategies and different organizational structures. But structure and strategy aren't enough.

This is where values come in. In a more volatile and dynamic business environment, the controls have to be conceptual. They can't be human anymore: Bob Haas telling people what to do. It's the *ideas* of a business that are controlling, not some manager with authority. Values provide a common language for aligning a company's leadership and its people.

Why isn't a sound business strategy enough to create that alignment?

A strategy is no good if people don't fundamentally believe in it. We had a strategy in the late 1970s and early 1980s that emphasized diversification. We acquired companies, created new brands, and applied our traditional brand to different kinds of apparel. Our people did what they were asked to do, but the problem was, they didn't believe in it.

The big change at Levi is that we have worked hard to listen to our suppliers, our customers, and our own people. We have redefined our business strategy to focus on core products, and we have articulated the values that the company stands for—what we call our Aspirations. We've reshaped our business around this strategy and these values, and people have started marching behind this new banner. In fact, they are running to grab it and take it on ahead of senior management. Because it's what they *want* to do.

At Levi, we talk about creating an "empowered" organization. By that, we mean a company where the people who are closest to the product and the customer take the initiative without having to check with anyone. Because in an organization of 31,000 people, there's no way that any one of us in management can be around all the time to tell people what to do. It has to be the strategy and the values that guide them.

What is the role of a manager in an empowered company?

If the people on the front line really are the keys to our success, then the manager's job is to help those people and the people that they serve. That goes against the traditional assumption that the manager is in control. In the past, a manager was expected to know everything that was going on and to be deeply involved in subordinates' activities.

I can speak from experience. It has been difficult for me to accept the fact that I don't have to be the smartest guy on the block—reading every memo and signing off on every decision. In reality, the more you establish parameters and encourage people to take initiatives within those boundaries, the more you multiply your own effectiveness by the effectiveness of other people.

So in a business world without boundaries, the chief role of managers is to establish some?

To set parameters. Those parameters are going to be different for different individuals. And for the same individual, they're going to be different for different tasks. Some people are going to be very inexperienced in certain things, so you need to be careful about setting the parameters of where they have authority and where they need to stop to seek clarification. Other people have experience, skills, and a track record, and within certain areas you want to give them a lot of latitude.

How does that compare with the traditional manager's job?

In many ways, it's a much tougher role because you can't rely on your title or unquestioning loyalty and obedience to get things done. You have to be thoughtful about what you want. You have to be clear about the standards that you're setting. You have to negotiate goals with your work group rather than just set them yourself. You have to interact personally with individuals whom you're dealing with, understand their strengths and shortcomings, and be clear about what you want them to do.

You also have to accept the fact that decisions or recommendations may be different from what you would do. They could very well be better, but they're going to be different. You have to be willing to take your ego out of it.

That doesn't mean abdication. Managers still have to make decisions, serve as counselors and coaches, be there when things get sticky, and help sort out all the tangles.

What else do managers in an empowered organization need to do?

They can clear away the obstacles to effective action that exist in any large organization. In most companies, including ours, there is a gap between what the organization says it wants and what it feels like to work there. Those gaps between what you say and what you do erode trust in the enterprise and in the leadership, and they inhibit action. The more you can narrow that gap, the more people's energies can be released toward company purposes.

Most people want to make a contribution and be proud of what they do. But organizations typically teach us bad habits—to cut corners, protect our own turf, be political. We've discovered that when people talk about what they want for themselves and for their company, it's very idealistic and deeply emotional. This company tells people that idealism is OK. And the power that releases is just unbelievable. Liberating those forces, getting the impediments out of the way, that's what we as managers are supposed to be doing.

What is happening in the apparel business that makes managing by values so important?

The same things that are happening in most businesses. For decades, apparel has been a very fragmented industry. Most producers were small. Typically, changes in manufacturing technology, the use of computers, and the application of marketing techniques came slowly if at all. Our customers were also highly fragmented. In the old days, we had some 18,000 domestic accounts of all sizes and in every town in the country.

In this environment, we considered ourselves a manufacturer. Our job was to design products, manufacture them, deliver them in accordance with our retailers' orders, and support the retailers with some consumer advertising to help the products sell. But the rest was up to the individual retailer.

Now all that is changing very rapidly. Today the top 50 accounts make up a large part of our domestic business. Style changes happen more rapidly because of innovations in fabric finishes and also because customers adopt new fashions more quickly. Technology is transforming sewing work and our relationships with our suppliers and customers. And we are operating in a global marketplace against international competition. As a result, the way we see our business has also changed.

What's the new vision?

First, we are a marketer rather than a manufacturer. And second, we are at the center of a seamless web of mutual responsibility and collaboration.

Take our relationships with our retailers: to secure the availability of a product, apparel retailers have traditionally had to order it as much as four to five months in advance. That's crazy. It forces retail buyers to guess four and five months down the road what a consumer who is 15 years old is going to want in jeans. During that time, a new movie can come out, and the trend goes from blue denim to black denim. And suddenly that inventory commitment is obsolete, causing costly markdowns.

One answer to this new circumstance is technology. Our electronic data-interchange system, LeviLink, was a pioneering effort in apparel to communicate with our customers and manage the order-replenishment cycle faster and more accurately than conventional systems could (see "How Values Shape Technology"). As a result, we have customers operating with 20% to 30% less inventory and achieving 20% to 30% increased sales. Their return on their investment with us is much greater than it was in the past. And these retailers also serve their customers better because the desired product is in stock when the consumer goes to purchase it.

How Values Shape Technology

According to Chief Information Officer Bill Eaton, the Levi Strauss & Co. technology strategy is a direct reflection of the company's Aspirations Statement. "Empowerment is meaningless unless people have access to information," says Eaton. "The goal of our technology strategy is to make sure that the information is available on the desktop of the person who is doing the job."

To that end, Levi Strauss has embarked on a three-part program for the global integration of its business through information technology. The most visible part of this strategy is LeviLink, the electronic data-interchange system that ties retailers to the company's distribution network. The system collects point-of-sale information from cash registers at the company's major accounts, then uses the information to generate reorders, invoices, packing slips, and advance notifications to retailers of future shipments. It also provides company sales representatives with far more information on the activity of individual retailers than was available in the past. Currently, about 40% of the company's business comes through LeviLink—a figure that the company hopes to double over the next five years.

Although less visible, Levi Strauss has also made major strides in computerizing its manufacturing operations. In 10 of the company's 32 factories worldwide, every sewing station now comes equipped with a hand-sized computer terminal. As a bundle of fabric moves through the plant, each employee who works on it passes the bundle's bar-coded label through a scanner built into the terminal. The result is a "real-time production control" system that allows the company to track work-in-process as it moves through the factory. The system also provides workers with information on their own performance, which they can use to increase productivity.

The ultimate goal is to link these two systems, allowing the company to issue production orders for new products immediately as existing products are sold in retailers' stores. This capacity will be provided by the Levi's Advanced Business System (LABS), which the company will be implementing over the next three years. LABS will be able to track a product from its conception—including orders, inventory, and financial information. Based on a new "relational database" software architecture, LABS will allow employees to perform more powerful and more flexible searches of company databases and to get rapid access to information typically limited in the past to managers.

The way Levi Strauss manages the development of new systems also reflects its commitment to using technology to support people. CIO Eaton serves on the company's executive management committee, ensuring the integration of technology and business strategies. And close ties between the information systems and human resources departments—one HR staff member now works in information systems—help connect the development of the company's new information technology platform to the creation of new business processes, organization designs, and people skills.

We're also forming closer relationships with our suppliers. We used to have ten or twelve U.S. denim suppliers. Now we're down to four or five. There is a seamless partnership, with interrelationships and mutual commitments, straight through the chain that would've been unimaginable ten years ago. You can't be responsive to the end-consumer today unless you can count on those kinds of collaborations at each step along the way.

What are the implications of that seamless partnership for your work force?

Our employees have many new responsibilities. For example, because of our computer linkages to our customers, our account representatives have more information on what's selling at the store level than the retailer does—not only products but sizes, fabrics, and styles. The rep has to know how to analyze that information and interpret it for the customer. What's more, since the computer does all the mundane record-keeping now, the rep can concentrate on planning and projecting the store's needs and being a marketing consultant.

What our employees do for the retailers is also much broader. In addition to the account representative, we have merchandising coordinators who make sure the stock is replenished and train salesclerks so they know how to sell our products more effectively. We have specialists in "visual merchandising" who work with our accounts to improve the ways they display our products. We have promotions experts who help stores tailor promotions to their clientele. And we run consumer hot lines to help customers find the products they want.

The work is much more creative, more entrepreneurial. It's as if these people are in business for themselves. They're doing what human beings do best—think, plan, interact, see trends, humanize the business to make it more successful.

What does that have to do with values?

To do that kind of work effectively requires a whole new set of attitudes and behaviors. The passivity and dependence of traditional paternalism—doing what you're told—doesn't work anymore. People have to take responsibility, exercise initiative, be accountable for their own success and for that of the company as a whole. They have to communicate more frequently and more effectively with their colleagues and their customers.

In a traditional command-and-control organization, acting in this way is difficult, even risky. The Aspirations encourage and support the new behaviors that we need. For example, in an empowered organization there are bound to be a lot more disagreements. Because we value open and direct communication, we give people permission to disagree. They can tell a manager, "It doesn't seem aspirational to be working with that contractor because from what we've seen, that company really mistreats its workers." Or they can say, "It may help us conserve cash to be slow in paying our bills, but that company has been a supplier for a long time, and it's struggling right now. Wouldn't it be better in terms of the partnership we're trying to create with our suppliers to pay our bills on time?"

Those are very challenging discussions for peers to have—let alone for somebody to have with his or her boss. But if we can "sanctify" it by reference to commonly held standards that we all share, it makes it all right to disagree.

So the values help bring about the kind of business behavior you need to stay competitive.

Values are where the hard stuff and the soft stuff come together. Let me give another example: in the new, more dynamic business environment, a company has to understand the relationship between work and family. It used to be that what happened to your employees when they went home at the end of the day was their business. But today that worker's sick child is your business, because if she's worrying about her child or calling in sick when she isn't—and probably feeling resentful because she's had to lie—she isn't going to be productive.

By contrast, if employees aren't worrying about things outside the workplace, if they feel supported—not just financially but "psychically"—then they are going to be more responsive to the needs of customers and of the business. That support needs to come in a whole set of managerial areas: supervisory practices, peer relations, training, work organization, access to information, and the like.

What is Levi doing about this particular issue?

We've established a companywide task force that's looking at how to balance work and family commitments. In itself, that's no big deal. A lot of companies are studying the issue. But even the way we manage the task force reflects our values. For instance, I'm on the task force, but I don't run it. We have everyone from secretaries and sewing machine operators to senior managers on the task force—as part of our commitment to diversity.

And that too makes perfect business sense. After all, my family situation is about as traditional as it gets. I have a wife at home who looks after our daughter. What do I know about the problems of a sewing machine operator—expected to punch in at a certain time and punch out at another and with a half-hour lunch break—whose child's day-care arrangements fall through that morning? Obviously, a better result is going to come out of a broad task force that represents a diversity of opinions, family situations, and points of view (see "The Making of an Aspiration").

The Making of an Aspiration

In 1985, a small group of minority and women managers asked for a private meeting with Levi Strauss & Co. CEO Robert Haas. The company had always been committed to equal employment opportunity; compared with other corporations, its numbers were good. But the managers felt strongly that they had reached a plateau. There were invisible barriers keeping women and minorities from advancing in the organization.

In response to their concerns, Haas organized an off-site retreat. Ten senior managers, all white men, were paired with either a woman or minority manager from their work group. The senior managers believed the company had been doing a good job hiring and promoting women and minorities. They were surprised to discover the depth of frustration and anger their subordinates felt. After two-and-a-half days of often painful discussion, the group concluded that equal opportunity was not just a matter of numbers but of attitudes, and that considerable unconscious discrimination still existed at the company. Something more needed to be done.

Since that 1985 meeting, Levi Strauss has renewed its commitment to full diversity at all levels. It has also broadened the definition of diversity beyond equal employment opportunity to the active encouragement of different points of view and their inclusion in company decision making.

Between 1985 and 1988, 16 more off-site sessions paired white, male managers with women or minorities who work for them to reflect on unexamined assumptions about diversity and stereotypes about particular groups. In 1987, diversity became one of the six Aspirations defined in the company's Aspirations Statement. That same year, members of the company's executive management committee—the top 8 people in the company—began holding monthly forums for small groups of 15 to 20 employees. By 1989, 20 ongoing forums were taking place. And while the original focus of the meetings was on questions of race and gender, the forums have expanded to consider a broad range of workplace issues.

The forums have been the catalyst for a number of new initiatives:

Recognizing that women and minority employees often have special problems and needs, the company has inaugurated four new career-development courses—one each for women, blacks, Hispanics, and Asians.

These courses have led to the creation of ethnic support networks for blacks, Asians, and Hispanics. Representatives of these networks have direct feedback to top management through quarterly meetings with the senior personnel committee.

In 1989, Levi Strauss established a companywide task force to recommend new policies to support a more effective balance between work and family life. Its 18 members are drawn from all levels of management and the work force—from sewing operators and secretaries to a division president and the CEO. This past summer, the task force sent a 25-page questionnaire to 15,000 of Levi's 21,000 U.S. employees to survey their family needs.

In 1989, the company also inaugurated a three-day course on "Valuing Diversity." About 240 top managers will have taken the course by the end of 1990. Eventually, all Levi Strauss employees will take diversity training.

How does a CEO manage for values?

The first responsibility for me and for my team is to examine critically our own behaviors and management styles in relation to the behaviors and values that we profess and to work to become more consistent with the values that we are articulating. It's tough work. We all fall off the wagon. But you can't be one thing and say another. People have unerring detection systems for fakes, and they won't put up with them. They won't put values into practice if you're not.

You said it's tough. What are the common kinds of breakdowns?

It's difficult to unlearn behaviors that made us successful in the past. Speaking rather than listening. Valuing people like yourself over people of different gender or from different cultures or parts of the organization. Doing things on your own rather than collaborating. Making the decision yourself instead of asking different people for their perspectives. There's a whole range of behaviors that were highly functional in the old hierarchical organization that are dead wrong in the flatter, more responsive, empowered organization that we're seeking to become.

Once your own behavior is in line with the new values, how do you communicate the values to others?

One way is to model the new behaviors we are looking for. For example, senior managers try to be explicit about our vulnerability and failings. We talk to people about the *bad* decisions we've made. It demystifies senior management and removes the stigma traditionally associated with taking risks. We also talk about the limitations of our own knowledge, mostly by inviting other people's perspectives.

When I talk to employees about the development of the Aspirations Statement [see below], I describe the stark terror I felt when I took over this company, just having turned 43 years old. We were a company in crisis. Our sales were dropping, our international business was heading for a loss, our domestic business had an eroding profit base, our diversification wasn't working, and we had too much production capacity. I had no bold plan of action. I knew that values were important but didn't have the two granite tablets that I could bring down from Mount Sinai to deliver to the organization. I talk about how alone I felt as a senior manager and how tough it is to be held up as a paragon to the organization. It helps people realize that senior management is human, that we can't be expected to know everything, and that we're inviting them in as partners in the organization.

Another important way to communicate values is through training. We've developed a comprehensive training program that we call the core curriculum. The centerpiece is a week-long course known as "leadership week" that helps managers practice the behavior outlined in the Aspirations Statement. We run about 20 sessions a year for a small group of about 20 people at a time. By the end of this year, the top 700 people in the company will have been through it. And at least one member of the executive management committee—the top 8 people in the company—or some other senior manager participates in every week-long session, just to send a signal of how important this is to us.

Aspirations Statement

We all want a company that our people are proud of and committed to, where all employees have an opportunity to contribute, learn, grow, and advance based on merit, not politics or background. We want our people to feel respected, treated fairly, listened to, and involved. Above all, we want satisfaction from accomplishments and friendships, balanced personal and professional lives, and to have fun in our endeavors.

When we describe the kind of Levi Strauss & Co. we want in the future, what we are talking about is building on the foundation we have inherited: affirming the best of our company's traditions, closing gaps that may exist between principles and practices, and updating some of our values to reflect contemporary circumstances.

What type of leadership is necessary to make our Aspirations a Reality?

New Behaviors: Leadership that exemplifies directness, openness to influence, commitment to the success of others, willingness to acknowledge our own contributions to problems, personal accountability, teamwork, and trust. Not only must we model these behaviors but we must coach others to adopt them.

Diversity: Leadership that values a diverse work force (age, sex, ethnic group, etc.) at all levels of the organization, diversity in experience, and diversity in perspectives. We have committed to taking full advantage of the rich backgrounds and abilities of all our people and to promoting a greater diversity in positions of influence. Differing points of view will be sought; diversity will be valued and honesty rewarded, not suppressed.

Recognition: Leadership that provides greater recognition—both financial and psychic—for individuals and teams that contribute to our success. Recognition must be given to all who contribute: those who create and innovate and also those who continually support the day-to-day business requirements.

Ethical Management Practices: Leadership that epitomizes the stated standards of ethical behavior. We must provide clarity about our expectations and must enforce these standards through the corporation.

Communications: Leadership that is clear about company, unit, and individual goals and performance. People must know what is expected of them and receive timely, honest feedback on their performance and career aspirations.

Empowerment: Leadership that increases the authority and responsibility of those closest to our products and customers. By actively pushing responsibility, trust, and recognition into the organization, we can harness and release the capabilities of all our people.

Can you really train people in new values?

You can't train anybody to do anything that he or she doesn't fundamentally believe in. That's why we've designed leadership week to give people an opportunity to reflect on their own values and to allow them to say what they want to get from work. In most cases, people learn that their personal values are aligned with those of the company. Of course, not everybody will buy into it. We've had some very honest discussions where managers say, "Look, I'm 53 years old, I've managed one way all my life and been successful, and now the company wants me to change. I don't know if I can do it."

But two things happen during leadership week. Because the groups are small, people build up a support network. They realize that others have the same problems that they have. Suddenly, they don't feel so alone.

Second, the training makes clear what's expected of them and what the consequences of succeeding or failing to adapt will be. It gives people the freedom to opt out. The real success of our core curriculum will be if it convinces some people that our environment is simply not right for them.

We also try to make sure that the core curriculum isn't just some nice experience that stops as soon as people get back to their jobs. For instance, there is a section of leadership week called "unanswered questions" where people voice concerns inspired by the course. Our human resources people collect these unanswered questions and report on them every quarter to the executive management committee. Sometimes, these questions can be handled by a particular individual. In other cases, we've set up a companywide task force to study the issue and come back with suggestions for changes in the way we do things. This creates a dialogue within the company among the people who have to make things happen.

Do the company's values influence the way you evaluate managers?

One-third of a manager's raise, bonus, and other financial rewards depends on his or her ability to manage aspirationally—the "how" of management rather than the "what." That goes for decisions about succession planning as well.

In some areas of the company, they're weighting it even more strongly. The point is, it's big enough to get people's attention. It's real. There's money attached to it. Giving people tough feedback and a low rating on aspirational management means improvement is necessary no matter how many pants they got out the door. Promotion is not in the future unless you improve.

How important is pay for pushing the company's values through the organization?

It's an influence but not the most important one. The key factors determining whether the values take or not will be individual commitment and desire and the peer pressure in the environment that we create. To me, the idea of a person as a marionette whose arms and legs start moving whenever you pull the pay string is too simplistic a notion of what motivates people in organizations.

That goes against the trend in recent years to fine-tune compensation incentives and tie them more closely to performance.

What happens there is you end up using pay to manage your company. But pay shouldn't manage your company; managers should. Managers should set the example, create the expectations, and provide the feedback. Managers should create an environment where people want to move in a constructive direction—not because there's money tied to the end of it, but because they feel it's right and they want to do it. That's why the way we conduct performance evaluations is probably more important than the pay we attach to aspirational management.

What's the process?

The typical performance evaluation in business has a manager set goals for a subordinate at the start of the year, and then at the end of the year make a subjective judgment about how well he or she has fulfilled them. That tends to create rabid "upward serving" behavior. People play to the person who's buttering their bread.

But what constitutes effective performance for a manager, anyway? Not necessarily pleasing your boss. Rather, good managers mobilize the talents of subordinates, peers, and clients to further the group's goals.

So what I started doing years ago—long before we developed the Aspirations Statement—was to talk with the direct reports of the people I manage, as well as with their peers and others they interact with. To evaluate one individual, I might interview anywhere from 10 to 16 people. The discussions are anonymous and confidential. And I report only trends, not isolated incidences.

This is an extremely powerful process that promotes ongoing feedback. The quotes may be anonymous, but they are very direct. "Here's what a group of people who work for you feel: You're much too controlling. You don't give them the latitude they need to show you how much they could do. People feel scared to take risks with you or to say controversial things because you act like you don't believe them."

I also ask the people who contribute to somebody's evaluation to say the same things to the person's face. If we can encourage regular dialogue among people so that they give their bosses or peers feedback on their performance, managers begin to realize that they have to pay attention to their people. If they create an unwholesome, unproductive environment and can't change it, we're not going to tolerate it.

Has the Aspirations Statement had any impact yet on the quality of major business decisions?

I'm the first to say our journey is incomplete. But compared with, say, five years ago, there definitely is a change. Suddenly, this $4 billion company feels like an owner-operated company, which is the goal.

Take the example of our Dockers product line. Dockers was like a new invention—a brand new segment in the casual pants market. The concept started in Argentina. Our Japanese affiliate picked it up under another name. Our menswear division adapted the idea to the U.S. market under the name of Dockers. Then the other domestic divisions saw its potential, and now it's in womenswear and kidswear. We also have a Shirts Dockers line and in selected international markets, we're seeing the startup of a Dockers product. In 1986, we sold 35,000 units. This year, we'll be doing a half-a-billion-dollar business in the United States.

We didn't have a business plan for Dockers. We had managers who saw an opportunity. They created a product and went out and made commitments for production that were greater than the orders they had in hand, because they believed in the product and its momentum. They got corporate support for an investment in advertising that was not justified on a cost-per-unit basis and created a product that anybody else in the market could have done. And five years later, it's a staple in the American wardrobe. None of this would have happened before this more collaborative, open style of management.

We've talked about getting managers to accept and live the company's values. But more than 75% of Levi's work force consists of operators in your sewing and finishing plants. Isn't the real challenge to make the company's values meaningful for them?

Empowerment isn't limited to just white-collar workers. By utilizing our people more fully than the apparel industry traditionally has, we can organize sewing work in ways that are much more in keeping with our Aspirations. About a year ago, we initiated an experiment at our Blue Ridge, Georgia plant, where we set up a gain-sharing program. We said to the employees, "You are the experts. If you meet predetermined production goals and predetermined absenteeism and safety standards, we'll split 50-50 with you any savings that result from economies or productivity improvements."

Sewing machine operators are now running the plant. They're making the rules and in some cases changing them because they understand why the rules are there and which rules make sense and which don't. They're taking initiatives and making things work better because it's in their interest and they don't have to be told.

This was not an unproductive plant. It was among the top 10% in the company. Today it's one of the top two plants—after only nine months in the new program. The financial payoff has been considerable, and there's certainly more potential there. But to me, the most exciting thing is to see the transformation in the workplace. People who felt that they weren't valued despite maybe 20 years of work for the company have a completely different attitude about their work.

In my judgment, we can restructure the workplace far beyond what we've done in Blue Ridge. I see us moving to a team-oriented, multiskilled environment in which the team itself takes on many of the supervisor's and trainer's tasks. If you combine that with some form of gain sharing, you probably will have a much more productive plant with higher employee satisfaction and commitment.

During the same period in which Levi has been defining its values, you've also been downsizing the company. Is it possible to get a high-commitment work force without offering some kind of employment security?

You can't promise employment security and be honest. The best you can do is not play games with people. You can't make any guarantees.

Through the 1950s, 1960s, and 1970s, Levi was growing so dramatically that unless you committed a felony, you weren't going to lose your job. But that was a special era stimulated by economic expansion and tremendous demographic growth. Now we're in a real-world situation where market forces are less favorable, external competitive pressures are more intense, and change is more rapid. You have to help people appreciate the need to deal constructively with the changing environment. If we're doing our job, we need to understand the rapidity and magnitude of the changes taking place and provide people with all the tools we can to cope with change.

But isn't it disingenuous to be championing values like empowerment in an environment where workers are worried about losing their jobs?

There is an apparent contradiction but not a real one, because our most basic value is honesty. If we have too much capacity, it's a problem that affects the entire company. Sometimes, the only solution is to close a plant, and if we don't have the guts to face that decision, then we risk hurting a lot of people—not just those in one plant. We need to be honest about that.

We tie it to Aspirations by asking, "How are we going to treat people who are displaced by technology, by changes in production sources, or by market changes?" We are committed to making the transition as successful as possible and to minimizing the uprooting and dislocation. We give more advance notice than is required by law. We provide more severance than is typical in our industry, so the effect of displacement is cushioned. We extend health care benefits. We also support job-training programs and other local initiatives to help our former employees find new jobs. And in the community itself, which has been depending on us as a major employer, we continue for a period of time to fund community organizations and social causes that we've been involved with, so that our withdrawal isn't a double hit—a loss of employment and also a loss of philanthropic support.

But has the Aspirations Statement changed the way you make decisions about capacity and plant closings in the first place?

The Aspirations make us slow down decisions. We challenge ourselves more explicitly to give some factors more weight than we did before—especially the impact of a plant closing on the community. There have been plants we have decided not to close, even though their costs were higher than other plants we did close. The reason was the community impact.

The Aspirations also provide a way to talk about these difficult trade-offs inside the company. People now have the freedom and authority to say, "Is it aspirational to be closing a plant when we're having a good year?" Or, "If we must close this plant, are we meeting our responsibilities to the employees and their community?" That forces us to be explicit about all the factors involved. It causes us to slow up, reflect, and be direct with one another about what's happening.

If the company's values cause you to slow up, doesn't that make it more difficult to respond to fast-changing markets?

Only if you assume it's still possible to separate the hard stuff from the soft stuff. Most managers say they want to optimize their business decisions. My personal philosophy is to suboptimize business decisions. Too often, optimizing really means taking only one dimension of a problem into account. Suboptimizing means looking at more than one factor and taking into account the interests and the needs of all the constituents. When you do that, suddenly the traditional hard values of business success and the nontraditional soft values relating to people start blending. The result is a better business decision—and it can still be done quickly if your employees understand the company's values and are empowered to take action without layers of review.

You mentioned collecting "unanswered questions" from employees about the role of values in the business. What's the most difficult for you to answer?

One of the most frequent things I hear is: "When the next downturn in the business happens, is top management going to remain committed to Aspirations?" The only answer to that one is, "Test us." We hope we won't have a downturn, but even if we do, I have no doubts about what management's commitment is. Only the experience of going through that kind of a situation, however, will convincingly demonstrate that commitment.

Where is that commitment to let values drive the business leading?

We've launched an irreversible process. Now we have to support the commitment that the Aspirations Statement is creating and be willing to deal with the tough issues that it raises. Two years ago, I gave a speech about the Aspirations at one of our worldwide management meetings. At the end, I held up the Aspirations Statement and ripped it to shreds. And I said, "I want each of you to throw away the Aspirations Statement and think about what you want for the company and what kind of person you want to be in the workplace and what kind of a legacy you want to leave behind. If the result happens to be the Aspirations, that's fine. But if it happens to be something else, the important thing is that you think deeply about who you are and what you stand for. I have enough confidence in your judgment and motivations that I'll go with whatever you come up with."

The point is, the Levi Strauss of the future is not going to be shaped by me or even by the Aspirations Statement. It's going to be shaped by our people and their actions, by the questions they ask and the responses we give, and by how this feeds into the way we run our business.

PART

IV

Getting from Here to There

1
Why Change Programs Don't Produce Change

Michael Beer, Russell A. Eisenstat, and Bert Spector

In the mid-1980s, the new CEO of a major international bank—call it U.S. Financial—announced a companywide change effort. Deregulation was posing serious competitive challenges—challenges to which the bank's traditional hierarchical organization was ill-suited to respond. The only solution was to change fundamentally how the company operated. And the place to begin was at the top.

The CEO held a retreat with his top 15 executives where they painstakingly reviewed the bank's purpose and culture. He published a mission statement and hired a new vice president for human resources from a company well-known for its excellence in managing people. And in a quick succession of moves, he established companywide programs to push change down through the organization: a new organizational structure, a performance appraisal system, a pay-for-performance compensation plan, training programs to turn managers into "change agents," and quarterly attitude surveys to chart the progress of the change effort.

As much as these steps sound like a textbook case in organizational transformation, there was one big problem: two years after the CEO launched the change program, virtually nothing in the way of actual changes in organizational behavior had occurred. What had gone wrong?

The answer is "everything." Every one of the assumptions the CEO made—about who should lead the change effort, what needed changing, and how to go about doing it—was wrong.

U.S. Financial's story reflects a common problem. Faced with changing markets and increased competition, more and more companies are struggling to re-establish their dominance, regain market share, and in some cases, ensure their survival. Many have come to understand that the key to competitive success is to transform the way they function. They are reducing reliance on managerial authority, formal rules and procedures, and narrow divisions of work. And they are creating teams, sharing information, and delegating responsibility and accountability far down the hierarchy. In effect, companies are moving from the hierarchical and bureaucratic model of organization that has characterized corporations since World War II to what we call the task-driven organization where what has to be done governs who works with whom and who leads.

But while senior managers understand the necessity of change to cope with new competitive realities, they often misunderstand what it takes to bring it about. They tend to share two assumptions with the CEO of U.S. Financial: that promulgating companywide programs—mission statements, "corporate culture" programs, training courses, quality circles, and new pay-for-performance systems—will transform organizations, and that employee behavior is changed by altering a company's formal structure and systems.

In a four-year study of organizational change at six large corporations, we found that exactly the opposite is true: the greatest obstacle to revitalization is the idea that it comes about through companywide change programs, particularly when a corporate staff group such as human resources sponsors them. We call this "the fallacy of programmatic change." Just as important, formal organization structure and systems cannot lead a corporate renewal process.

While in some companies, wave after wave of programs rolled across the landscape with little positive impact, in others, more successful transformations did take place. They usually started at the periphery of the corporation in a few plants and divisions far from corporate headquarters. And they were led by the general managers of those units, not by the CEO or corporate staff people.

The general managers did not focus on formal structures and systems; they created ad hoc organizational arrangements to solve concrete business problems. By aligning employee roles, responsibilities, and relationships to address the organization's most important competitive task—a process we call "task alignment"—they focused energy for change on the work itself, not on abstractions such as "participation" or "culture." Unlike the CEO at U.S. Financial, they didn't employ massive training programs or rely on speeches and mission statements. Instead, we saw that general managers carefully developed the change process through a sequence of six basic managerial interventions.

Once general managers understand the logic of this sequence, they don't have to wait for senior management to start a process of organizational revitalization. There is a lot they can do even without support from the top. Of course, having a CEO or other senior managers who are committed to change does make a difference—and when it comes to changing an entire organization, such support is essential. But top management's role in the change process is very different from that which the CEO played at U.S. Financial.

Grass-roots change presents senior managers with a paradox: directing a "nondirective" change process. The most effective senior managers in our study recognized their limited power to mandate corporate renewal from the top. Instead, they defined their roles as creating a climate for change, then spreading the lessons of both successes and failures. Put another way, they specified the general direction in which the company should move without insisting on specific solutions.

In the early phases of a companywide change process, any senior manager can play this role. Once grass-roots change reaches a critical mass, however, the CEO has to be ready to transform his or her own work unit as well—the top team composed of key business heads and corporate staff heads. At this point, the company's structure and systems must be put into alignment with the new management practices that have developed at the periphery. Otherwise, the tension between dynamic units and static top management will cause the change process to break down.

We believe that an approach to change based on task alignment, starting at the periphery and moving steadily toward the corporate core, is the most effective way to achieve enduring organizational change. This is not to say that change can *never* start at the top, but it is uncommon and too risky as a deliberate strategy. Change is about learning. It is a rare CEO who knows in advance the fine-grained details of organizational change that the many diverse units of a large corporation demand. Moreover, most of today's senior executives developed in an era in which top-down hierarchy was the primary means for organizing and managing. They must learn from innovative approaches coming from younger unit managers closer to the action.

The Fallacy of Programmatic Change

Most change programs don't work because they are guided by a theory of change that is fundamentally flawed. The common belief is that the place to begin is with the knowledge and attitudes of individuals. Changes in attitudes, the theory goes, lead to changes in individual behavior. And changes in individual behavior, repeated by many people, will result in organizational change. According to this model, change is like a conversion experience. Once people "get religion," changes in their behavior will surely follow.

This theory gets the change process exactly backward. In fact, individual behavior is powerfully shaped by the organizational roles that people play. The most effective way to change behavior, therefore, is to put people into a new organizational context, which imposes new roles, responsibilities, and relationships on them. This creates a situation that, in a sense, "forces" new attitudes and behaviors on people. (See Exhibit I.)

One way to think about this challenge is in terms of three interrelated factors required for corporate revitalization. *Coordination* or teamwork is especially important if an organization is to discover and act on cost, quality, and product development opportunities. The production and sale of innovative, high-quality, low-cost products (or services) depend on close coordination among marketing, product design, and manufacturing departments, as well as between labor and management. High levels of *commitment* are essential for the effort, initiative, and cooperation that coordinated action demands. New *competencies* such as knowledge of the business as a whole, analytical skills, and interpersonal skills are necessary if people are to identify and solve problems as a team. If any of these elements are missing, the change process will break down.

The problem with most companywide change programs is that they address only one or, at best, two of these factors. Just because a company issues a philosophy statement about teamwork doesn't mean its employees necessarily know what teams to form or how to function within them to improve coordination. A corporate reorganization may change the boxes on a formal organization chart but not provide the necessary attitudes and skills to make the new structure work. A pay-for-performance system may force managers to differentiate better performers from poorer ones, but it doesn't help them internalize new standards by which to judge subordinates' performances. Nor does it teach them how to deal effectively with performance problems. Such programs cannot provide the cultural context (role models from whom to learn) that people need to develop new competencies, so ultimately they fail to create organizational change.

Similarly, training programs may target competence, but rarely do they change a company's patterns of coordination. Indeed, the excite-

Exhibit I. Contrasting Assumptions About Change

Programmatic Change	Task Alignment
Problems in behavior are a function of individual knowledge, attitudes, and beliefs.	Individual knowledge, attitudes, and beliefs are shaped by recurring patterns of behavioral interactions.
The primary target of renewal should be the content of attitudes and ideas; actual behavior should be secondary.	The primary target of renewal should be behavior; attitudes and ideas should be secondary.
Behavior can be isolated and changed individually.	Problems in behavior come from a circular pattern, but the effects of the organizational system on the individual are greater than those of the individual on the system.
The target for renewal should be at the individual level.	The target for renewal should be at the level of roles, responsibilities, and relationships.

ment engendered in a good corporate training program frequently leads to increased frustration when employees get back on the job only to see their new skills go unused in an organization in which nothing else has changed. People end up seeing training as a waste of time, which undermines whatever commitment to change a program may have roused in the first place.

When one program doesn't work, senior managers, like the CEO at U.S. Financial, often try another, instituting a rapid progression of programs. But this only exacerbates the problem. Because they are designed to cover everyone and everything, programs end up covering nobody and nothing particularly well. They are so general and standardized that they don't speak to the day-to-day realities of particular units. Buzzwords like "quality," "participation," "excellence," "empowerment," and "leadership" become a substitute for a detailed understanding of the business.

And all these change programs also undermine the credibility of the change effort. Even when managers accept the potential value of a particular program for others—quality circles, for example, to solve a manufacturing problem—they may be confronted with another, more pressing business problem such as new product development. One-size-fits-all change programs take energy *away* from efforts to solve key business problems—which explains why so many general managers don't support programs, even when they acknowledge that their underlying principles may be useful.

This is not to state that training, changes in pay systems or organizational structure, or a new corporate philosophy are always inappropriate. All can play valuable roles in supporting an integrated change effort. The problems come when such programs are used in isolation as a kind of "magic bullet" to spread organizational change rapidly through the entire corporation. At their best, change programs of this sort are irrelevant. At their worst, they actually inhibit change. By promoting skepticism and cynicism, programmatic change can inoculate companies against the real thing.

Six Steps to Effective Change

Companies avoid the shortcomings of programmatic change by concentrating on "task alignment"—reorganizing employee roles, responsibilities, and relationships to solve specific business problems. Task alignment is easiest in small units—a plant, department, or business unit—where goals and tasks are clearly defined. Thus the chief problem for corporate change is how to promote task-aligned change across many diverse units.

We saw that general managers at the business unit or plant level can achieve task alignment through a sequence of six overlapping but distinctive steps, which we call the *critical path*. This path develops a self-reinforcing cycle of commitment, coordination, and competence. The sequence of steps is important because activities appropriate at one time are often counterproductive if started too early. Timing is everything in the management of change.

1. Mobilize commitment to change through joint diagnosis of business problems. As the term task alignment suggests, the starting point of any effective change effort is a clearly defined business problem. By helping people develop a shared diagnosis of what is wrong in an organization and what can and must be improved, a general manager mobilizes the initial commitment that is necessary to begin the change process.

Consider the case of a division we call Navigation Devices, a business unit of about 600 people set up by a large corporation to commercialize a product originally designed for the military market. When the new general manager took over, the division had been in operation for several years without ever making a profit. It had never been able to design and produce a high-quality, cost-competitive product. This was due largely to an organization in which decisions were made at the top, without proper involvement of or coordination with other functions.

The first step the new general manager took was to initiate a broad review of the business. Where the previous general manager had set strategy with the unit's marketing director alone, the new general manager included his entire management team. He also brought in outside consultants to help him and his managers function more effectively as a group.

Next, he formed a 20-person task force representing all the stakeholders in the organization—managers, engineers, production workers, and union officials. The group visited a number of successful manufacturing organizations in an attempt to identify what Navigation Devices might do to organize more effectively. One high-performance manufacturing plant in the task force's own company made a particularly strong impression. Not only did it highlight the problems at Navigation Devices but it also offered an alternative organizational model, based on teams, that captured the group's imagination. Seeing a different way of working helped strengthen the group's commitment to change.

The Navigation Devices task force didn't learn new facts from this process of joint diagnosis; everyone already knew the unit was losing money. But the group came to see clearly the organizational roots of the unit's inability to compete and, even more important, came to share a common understanding of the problem. The group also identified a potential organizational solution: to redesign the way it worked, using ad hoc teams to integrate the organization around the competitive task.

2. Develop a shared vision of how to organize and manage for competitiveness. Once a core group of people is committed to a particular analysis of the problem, the general manager can lead employees toward a task-aligned vision of the organization that defines new roles and responsibilities. These new arrangements will coordinate the flow of information and work across interdependent functions at all levels of the organization. But since they do not change formal structures and systems like titles or compensation, they encounter less resistance.

At Navigation Devices, the 20-person task force became the vehicle for this second stage. The group came up with a model of the organization in which cross-functional teams would accomplish all work, particularly new product development. A business-management team composed of the general manager and his staff would set the unit's strategic direction and review the work of lower level teams. Business-area teams would develop plans for specific markets. Product-development teams would manage new products from initial design to production. Production-process teams composed of engineers and production workers would identify and solve quality and cost problems in the plant. Finally, engineering-process teams would examine engineering methods and equipment. The teams got to the root of the unit's problems—functional and hierarchical barriers to sharing information and solving problems.

To create a consensus around the new vision, the general manager commissioned a still larger task force of about 90 employees from different levels and functions, including union and management, to refine the vision and obtain everyone's commitment to it. On a retreat away from the workplace, the group further refined the new organizational model and drafted a values statement, which it presented later to the entire Navigation Devices work force. The vision and the values statement made sense to Navigation Devices employees in a way many corporate mission statements never do—because it grew out of the organization's own analysis of real business problems. And it was built on a model for solving those problems that key stakeholders believed would work.

3. Foster consensus for the new vision, competence to enact it, and cohesion to move it along. Simply letting employees help develop a new vision is not enough to overcome resistance to change—or to foster the skills needed to make the new organization work. Not everyone can help in the design, and even those who do participate often do not fully appreciate what renewal will require until the new organization is actually in place. This is when strong leadership from the general manager is crucial. Commitment to change is always uneven. Some managers are enthusiastic; others are neutral or even antagonistic. At Navigation Devices, the general manager used what his subordinates termed the "velvet glove." He made it clear that the division was going to encourage employee involvement and the team approach. To managers who wanted to help him, he offered support. To those who did not, he offered outplacement and counseling.

Once an organization has defined new roles and responsibilities, people need to develop the competencies to make the new setup work. Actually, the very existence of the teams with their new goals and accountabilities will force learning. The changes in roles, responsibilities, and relationships foster new skills and attitudes. Changed patterns of coordination will also increase employee participation, collaboration, and information sharing.

But management also has to provide the right supports. At Navigation Devices, six resource people—three from the unit's human resources department and three from corporate headquarters—worked on the change project. Each team was assigned one internal consultant, who attended every meeting, to help people be effective team members. Once employees could see exactly what kinds of new skills they needed, they asked for formal training programs to develop those skills further. Since these courses grew directly out of the employees' own experiences, they were far more focused and useful than traditional training programs.

Some people, of course, just cannot or will not change, despite all the direction and support in the world. Step three is the appropriate time to replace those managers who cannot function in the new organization—after they have had a chance to prove themselves. Such decisions are rarely easy, and sometimes those people who have difficulty working in a participatory organization have extremely valuable specialized skills. Replacing them early in the change process, before they have worked in the new organization, is not only unfair to individuals; it can be demoralizing to the entire organization and can disrupt the change process. People's understanding of what kind of manager and worker the new organization demands grows slowly and only from the experience of seeing some individuals succeed and others fail.

Once employees have bought into a vision of what's necessary and have some understanding of what the new organization requires, they can accept the necessity of replacing or moving people who don't make the transition to the new way of working. Sometimes people are transferred to other parts of the company where technical expertise rather than the new competencies is the main requirement. When no alternatives exist, sometimes they leave the company through early retirement programs, for example. The act of replacing people can actually reinforce the organization's commitment to change by visibly demonstrating the general manager's commitment to the new way.

Some of the managers replaced at Navigation Devices were high up in the organization—for example, the vice president of operations, who oversaw the engineering and manufacturing departments. The new head of manufacturing was far more committed to change and skilled in leading a critical path change process. The result was speedier change throughout the manufacturing function.

4. Spread revitalization to all departments without pushing it from the top.

With the new ad hoc organization for the unit in place, it is time to turn to the functional and staff departments that must interact with it. Members of teams cannot be effective unless the department from which they come is organized and managed in a way that supports their roles as full-fledged participants in team decisions. What this often means is that these departments will have to rethink their roles and authority in the organization.

At Navigation Devices, this process was seen most clearly in the engineering department. Production department managers were the most enthusiastic about the change effort; engineering managers were more hesitant. Engineering had always been king at Navigation Devices; engineers designed products to the military's specifications without much concern about whether manufacturing could easily build them or not. Once the new team structure was in place, however, engineers had to participate on product-development teams with production workers. This required them to re-examine their roles and rethink their approaches to organizing and managing their own department.

The impulse of many general managers faced with such a situation would be to force the issue—to announce, for example, that now all parts of the organization must manage by teams. The temptation to force newfound insights on the rest of the organization can be great, particularly when rapid change is needed, but it would be the same mistake that senior managers make when they try to push programmatic change throughout a company. It short-circuits the change process.

It's better to let each department "reinvent the wheel"—that is, to find its own way to the new organization. At Navigation Devices, each department was allowed to take the general concepts of coordination and teamwork and apply them to its particular situation. Engineering spent nearly a year agonizing over how to implement the team concept. The department conducted two surveys, held off-site meetings, and proposed, rejected, then accepted a matrix management structure before it finally got on board. Engineering's decision to move to matrix management was not surprising, but because it was its own choice, people committed themselves to learning the necessary new skills and attitudes.

5. Institutionalize revitalization through formal policies, systems, and structures. There comes a point where general managers have to consider how to institutionalize change so that the process continues even after they've moved on to other responsibilities. Step five is the time: the new approach has become entrenched, the right people are in place, and the team organization is up and running. Enacting changes in structures and systems any earlier tends to backfire. Take information systems. Creating a team structure means new information require- ments. Why not have the MIS department create new systems that cut across traditional functional and departmental lines early in the change process? The problem is that without a well-developed under- standing of information requirements, which can best be obtained by placing people on task-aligned teams, managers are likely to resist new systems as an imposition by the MIS department. Newly formed teams can often pull together enough information to get their work done without fancy new systems. It's better to hold off until everyone understands what the team's information needs are.

What's true for information systems is even more true for other formal structures and systems. Any formal system is going to have some disadvantages; none is perfect. These imperfections can be mini- mized, however, once people have worked in an ad hoc team structure and learned what interdependencies are necessary. Then employees will commit to them too.

Again, Navigation Devices is a good example. The revitalization of the unit was highly successful. Employees changed how they saw their roles and responsibilities and became convinced that change could actually make a difference. As a result, there were dramatic improve- ments in value added per employee, scrap reduction, quality, customer service, gross inventory per employee, and profits. And all this hap- pened with almost no formal changes in reporting relationships, infor- mation systems, evaluation procedures, compensation, or control sys- tems.

When the opportunity arose, the general manager eventually did make some changes in the formal organization. For example, when he moved the vice president of operations out of the organization, he eliminated the position altogether. Engineering and manufacturing reported directly to him from that point on. For the most part, how- ever, the changes in performance at Navigation Devices were sus- tained by the general manager's expectations and the new norms for behavior.

6. Monitor and adjust strategies in response to problems in the revitalization process. The purpose of change is to create an asset that did not exist before—a learning organization capable of adapting to a changing competitive environment. The organization has to know how to con- tinually monitor its behavior—in effect, to learn how to learn.

Some might say that this is the general manager's responsibility. But monitoring the change process needs to be shared, just as analyzing the organization's key business problem does.

At Navigation Devices, the general manager introduced several mechanisms to allow key constituents to help monitor the revitalization. An oversight team—composed of some crucial managers, a union leader, a secretary, an engineer, and an analyst from finance—kept continual watch over the process. Regular employee attitude surveys monitored behavior patterns. Planning teams were formed and reformed in response to new challenges. All these mechanisms created a long-term capacity for continual adaptation and learning.

The six-step process provides a way to elicit renewal without imposing it. When stakeholders become committed to a vision, they are willing to accept a new pattern of management—here the ad hoc team structure—that demands changes in their behavior. And as the employees discover that the new approach is more effective (which will happen only if the vision aligns with the core task), they have to grapple with personal and organizational changes they might otherwise resist. Finally, as improved coordination helps solve relevant problems, it will reinforce team behavior and produce a desire to learn new skills. This learning enhances effectiveness even further and results in an even stronger commitment to change. This mutually reinforcing cycle of improvements in commitment, coordination, and competence creates a growing sense of efficacy. It can continue as long as the ad hoc team structure is allowed to expand its role in running the business.

The Role of Top Management

To change an entire corporation, the change process we have described must be applied over and over again in many plants, branches, departments, and divisions. Orchestrating this companywide change process is the first responsibility of senior management. Doing so successfully requires a delicate balance. Without explicit efforts by top management to promote conditions for change in individual units, only a few plants or divisions will attempt change, and those that do will remain isolated. The best senior manager leaders we studied held their subordinates responsible for starting a change process without specifying a particular approach.

Create a market for change. The most effective approach is to set demanding standards for all operations and then hold managers accountable to them. At our best-practice company, which we call General Products, senior managers developed ambitious product and operating standards. General managers unable to meet these product standards by a certain date had to scrap their products and take a sharp hit to their bottom lines. As long as managers understand that high standards are not arbitrary but are dictated by competitive forces, standards can generate enormous pressure for better performance, a key ingredient in mobilizing energy for change.

But merely increasing demands is not enough. Under pressure, most managers will seek to improve business performance by doing more of what they have always done—overmanage—rather than alter the fundamental way they organize. So, while senior managers increase demands, they should also hold managers accountable for fundamental changes in the way they use human resources.

For example, when plant managers at General Products complained about the impossibility of meeting new business standards, senior managers pointed them to the corporate organization-development department within human resources and emphasized that the plant managers would be held accountable for moving revitalization along. Thus top management had created a demand system for help with the new way of managing, and the human resources staff could support change without appearing to push a program.

Use successfully revitalized units as organizational models for the entire company. Another important strategy is to focus the company's attention on plants and divisions that have already begun experimenting with management innovations. These units become developmental laboratories for further innovation.

There are two ground rules for identifying such models. First, innovative units need support. They need the best managers to lead them, and they need adequate resources—for instance, skilled human resources people and external consultants. In the most successful companies that we studied, senior managers saw it as their responsibility to make resources available to leading-edge units. They did not leave it to the human resources function.

Second, because resources are always limited and the costs of failure high, it is crucial to identify those units with the likeliest chance of success. Successful management innovations can appear to be failures when the bottom line is devastated by environmental factors beyond the unit's control. The best models are in healthy markets.

Obviously, organizational models can serve as catalysts for change only if others are aware of their existence and are encouraged to learn from them. Many of our worst-practice companies had plants and divisions that were making substantial changes. The problem was, nobody knew about them. Corporate management had never bothered to highlight them as examples to follow. In the leading companies, visits, conferences, and educational programs facilitated learning from model units.

Develop career paths that encourage leadership development. Without strong leaders, units cannot make the necessary organizational changes, yet the scarcest resource available for revitalizing corporations is leadership. Corporate renewal depends as much on developing effective change leaders as it does on developing effective organizations. The personal learning associated with leadership development— or the realization by higher management that a manager does not have this capacity—cannot occur in the classroom. It only happens in an organization where the teamwork, high commitment, and new competencies we have discussed are already the norm.

The only way to develop the kind of leaders a changing organization needs is to make leadership an important criterion for promotion, and then manage people's careers to develop it. At our best-practice companies, managers were moved from job to job and from organization to organization based on their learning needs, not on their position in the hierarchy. Successful leaders were assigned to units that had been targeted for change. People who needed to sharpen their leadership skills were moved into the company's model units where those skills would be demanded and therefore learned. In effect, top management used leading-edge units as hothouses to develop revitalization leaders.

But what about the top management team itself? How important is it for the CEO and his or her direct reports to practice what they preach? It is not surprising—indeed, it's predictable—that in the early years of a corporate change effort, top managers' actions are often not consistent with their words. Such inconsistencies don't pose a major barrier to corporate change in the beginning, though consistency is obviously desirable. Senior managers can create a climate for grass-roots change without paying much attention to how they themselves operate and manage. And unit managers will tolerate this inconsistency so long as they can freely make changes in their own units in order to compete more effectively.

There comes a point, however, when addressing the inconsistencies becomes crucial. As the change process spreads, general managers in the ever-growing circle of revitalized units eventually demand changes from corporate staff groups and top management. As they discover how to manage differently in their own units, they bump up against constraints of policies and practices that corporate staff and top management have created. They also begin to see opportunities for better coordination between themselves and other parts of the company over which they have little control. At this point, corporate organization must be aligned with corporate strategy, and coordination between related but hitherto independent businesses improved for the benefit of the whole corporation.

None of the companies we studied had reached this "moment of truth." Even when corporate leaders intellectually understood the direction of change, they were just beginning to struggle with how they would change themselves and the company as a whole for a total corporate revitalization.

This last step in the process of corporate renewal is probably the most important. If the CEO and his or her management team do not ultimately apply to themselves what they have been encouraging their general managers to do, then the whole process can break down. The time to tackle the tough challenge of transforming companywide systems and structures comes finally at the end of the corporate change process.

At this point, senior managers must make an effort to adopt the team behavior, attitudes, and skills that they have demanded of others in earlier phases of change. Their struggle with behavior change will help sustain corporate renewal in three ways. It will promote the attitudes and behavior needed to coordinate diverse activities in the company; it will lend credibility to top management's continued espousal of change; and it will help the CEO identify and develop a successor who is capable of learning the new behaviors. Only such a manager can lead a corporation that can renew itself continually as competitive forces change.

Companies need a particular mind-set for managing change: one that emphasizes process over specific content, recognizes organization change as a unit-by-unit learning process rather than a series of programs, and acknowledges the payoffs that result from persistence over a long period of time as opposed to quick fixes. This mind-set is difficult to maintain in an environment that presses for quarterly earnings, but we believe it is the only approach that will bring about successful renewal.

2

Motorola U: When Training Becomes an Education

William Wiggenhorn

At Motorola we require three things of our manufacturing employees. They must have communication and computation skills at the seventh grade level, soon going up to eighth and ninth. They must be able to do basic problem solving—not only as individuals but also as members of a team. And they must accept our definition of work and the workweek: the time it takes to ship perfect product to the customer who's ordered it. That can mean a workweek of 50 or even 60 hours, but we need people willing to work against quality and output instead of a time clock.

These requirements are relatively new. Ten years ago, we hired people to perform set tasks and didn't ask them to do a lot of thinking. If a machine went down, workers raised their hands, and a troubleshooter came to fix it. Ten years ago, we saw quality control as a screening process, catching defects before they got out the door. Ten years ago, most workers and some managers learned their jobs by observation, experience, and trial and error. When we did train people, we simply taught them new techniques on top of the basic math and communication skills we supposed they brought with them from school or college.

Then all the rules of manufacturing and competition changed, and in our drive to change with them, we found we had to rewrite the rules of corporate training and education. We learned that line workers had to actually understand their work and their equipment, that senior management had to exemplify and reinforce new methods and skills if they were going to stick, that change had to be continuous and participative, and that education—not just instruction—was the only way to make all this occur.

Finally, just as we began to capitalize on the change we thought we were achieving, we discovered to our utter astonishment that much of our work force was illiterate. They couldn't read. They couldn't do simple arithmetic like percentages and fractions. At one plant, a supplier changed its packaging, and we found in the nick of time that our people were working by the color of the package, not by what it said. In Illinois, we found a foreign-born employee who didn't know the difference between the present tense and the past. He was never sure if we were talking about what *was* happening or what *had* happened.

These discoveries led us into areas of education we had never meant to enter and into budgetary realms we would have found unthinkable ten years earlier. From the kind of skill instruction we envisioned at the outset, we moved out in both directions: down, toward grade school basics as fundamental as the three Rs; up, toward new concepts of work, quality, community, learning, and leadership. From a contemplated total budget of $35 million over a five-year period, a sum many thought excessive, we came to spend $60 million annually—plus another $60 million in lost work time—and everyone thought it was money well invested.

Today we expect workers to know their equipment and begin any troubleshooting process themselves. If they do need an expert, they must be able to describe the malfunction in detail. In other words, they have to be able to analyze problems and then communicate them.

Today we see quality as a process that prevents defects from occurring, a common corporate language that pervades the company and applies to security guards and secretaries as well as manufacturing staff. (See "The Language of Quality.")

The Language of Quality

The mathematics of quality are difficult. Even the vocabulary—bell curves, probability functions, standard deviations expressed in multiples of the Greek letter Σ—can be formidable. At Motorola, we have nevertheless tried to teach at least a basic version of this math to every employee and to extend the concepts and terminology of industrial quality into every corner of the business—training, public relations, finance, security, even cooking. In 1983, we thought three days of training on quality would be enough. Today we have 28 days of material—quality tools, quality strategy, quality techniques, quality feedback mechanisms—and over the course of several years, we expect our engineers and manufacturing middle managers to have all 28.

The corporate goal is to achieve a quality standard by 1992 equivalent to what the statisticians and industrial engineers call Six Sigma, which means six standard deviations from a statistical performance average. In plain English, Six Sigma translates into 3.4 defects per million opportunities, or production that is 99.99966% defect free. (By contrast, and according to formulas not worth explaining here, 5Σ is 233 defects per million, and 4Σ is 6,210. Airlines achieve 6.5Σ in safety—counting fatalities as defects—but only 3.5 to 4Σ in baggage handling. Doctors and pharmacists achieve an accuracy of just under 5Σ in writing and filling prescriptions.)

Motorola has not yet reached Six Sigma in manufacturing or in any other function. But we have achieved our initial goal of creating a common vocabulary that lets every person in the company speak the same quality language. Everyone pursues his or her own job version of Six Sigma, and everyone shares a sense of what it means, objectively and subjectively, to take part in the process of getting from where we are to where we want to be. The idea of 3.4 defects per million opportunities may sound ridiculous for a course instructor or a chef, but there is always some way to apply the standard and strive for the goal. In effect, the Six Sigma process means changing the way people do things so that nothing can go wrong.

Applied to the chef, the process of reaching Six Sigma by 1992 means he can burn five muffins this year, two muffins next year, and eventually none at all. Of course, Six Sigma applied to muffins is in one sense a fiction, and if we presented it to the chef as an ultimatum, it would also be insulting. The real point is that importing that quality language from manufacturing to the rest of the company stimulates a kind of discussion we might not otherwise have had. It also tells people they're important, since the time-cycle and quality standards so vital to manufacturing and product design now also apply to the chef, the security guard, and the clerical support people.

What we actually said to the chef was that he made a consistently Six Sigma chocolate chip cookie—in my opinion it ranks with the best in the world—and we wanted him to do the same with muffins. We asked him what changes we needed to make so he could accomplish that. He said, "The reason I can't make better muffins is that you don't trust me with the key to the freezer. I have to make the batter the night before because when I get here in the morning the makings are still locked up. When batter sits for 12 hours, it's never as good as batter you make fresh and pop directly in the oven." He was right, of course. The language of quality can also be used to talk about trust.

Security guards are another good example. Every morning they have to get 12,000 employees and about 400 visitors through the gates in half an hour. What's their assignment—to keep people out, bring people in, protect property? As we now define it, their job is to make sure the right people get in quickly and, if there are problems, to handle them politely and professionally. Quality is measured accordingly—a "customer" complaint is a defect, and a customer is anyone who comes to the gate.

Today Motorola has one of the most comprehensive and effective corporate training and education programs in the world and, in a recent leap of ambition, our own corporate university.

Why in the world, you ask, should any corporation have its own university? My answer is the story of how we came to have ours. In part, it is a kind of odyssey, a ten-year expedition full of well-meaning mistakes, heroic misapprehensions, and shocking discoveries. In part, it is a slowly unfolding definition of education and change that shows why successful companies in today's business climate must not only train workers but build educational systems.

An MBA in Four Weeks

In 1979, Bob Galvin, then Motorola's CEO and now chairman of the executive committee, asked the human resources department to put together a five-year training plan. He believed that all employees needed upgrading in their skills if the company was going to survive.

Galvin had made two earlier attempts at companywide education. The first focused on new tools, new technology, and teamwork, but it didn't produce the results he wanted. Plant managers brought in new equipment but wouldn't change the support systems and their own work patterns.

So he set up the Motorola Executive Institute, an intensive, one-time course for 400 executives that tried to give them an MBA in four weeks. The participants learned a great deal, but again, the ultimate results were disappointing.

Galvin, who understood that change had to force its way through a company from the top down, had been driving home the point that those who lead often lose their power—or their right to lead—because they're unwilling to change. He now realized that the top probably wasn't going to lead the attack until all employees wanted change to take place. Motorola had to educate everyone and make people see the need for change.

To carry out this training program, we set up an education service department—MTEC, the Motorola Training and Education Center—with its own board of directors consisting of Galvin himself, two of his top executives, and senior managers from each of Motorola's operating units. MTEC had two principal goals: to expand the participative management process and to help improve product quality tenfold in five years.

Our charter was not so much to educate people as to be an agent of change, with an emphasis on retraining workers and redefining jobs. Our first order of business was to analyze the jobs that existed then, in 1980, and try to anticipate what they'd look like in the future. The first thing we learned was not to look too far ahead. If we made a two-year projection and trained people for that, then change didn't arrive quickly enough for people to make the shift. We had to anticipate, plan curricula, then train separately for each incremental change. We had thought progress would be made in leaps, but it took place one step at a time.

To meet the quality target, we developed a five-part curriculum. First came statistical process control, which consisted of instruction in seven quality tools. Basic industrial problem solving was second. Third was a course on how to present conceptual material, a tricky assignment for an hourly worker presenting a technical solution to an engineer. Fourth was a course on effective meetings that emphasized the role of participant as well as that of chairperson. Finally we had a program on goal setting that taught people how to define objectives, how to describe them in writing, and how to measure progress.

So far so good. In the early 1980s, at a typical plant with 2,500 workers, MTEC was using 50,000 hours of employee time—a lot of time away from the job for what some people considered a pretty esoteric program. We thought it was worth the investment. Putting quality tools in the hands of every employee was the only way to overcome the old emphasis on shipment goals, even when meeting those goals meant shipping defective products.

Yet the skeptics were right. We were wasting everyone's time. We designed and taught courses, and people took them and went back to their jobs, and nothing changed. We had made a series of false assumptions.

Getting People to Want to Learn

Our first mistake was to assume that once we described the courses, the people who needed them most would sign up to take them. They didn't. We also assumed that the courses would be popular, but enrollment was never in danger of swamping our capacities.

The old approach had been to learn by watching others. When technology changed once in five years, on-the-job training made some sense, but people can't handle constant innovation by watching one another. Yet somehow the culture told them they didn't learn things any other way. Since people resisted formal classes, we developed self-help material so they could pick up a package and take it home. That failed too. People just didn't see homework as real training, which left us in a bind: our employees didn't seem to believe the training was necessary, but if it was necessary, then it had to take place in a formal classroom, not at home. So we dropped the learn-at-home program. Not because people couldn't learn that way but because we couldn't get people to *want* to learn that way.

Training, it appeared, was not something we could deliver like milk and expect people to consume spontaneously. It was not simply a matter of instructing or giving people a chance to instruct themselves. We had to motivate people to want to learn, and that meant overcoming complacency.

When Motorola hired people in the old days, we hired them for life. People grew up in their jobs, acquired competencies and titles, moved from work force to management. All employees became members of our Service Club at the end of ten years, which meant we wouldn't terminate them except for poor performance or dishonesty. We never gave anyone an absolute right to lifelong employment, but we did provide an unmistakable opportunity to stay.

This was the employment model that built the corporation and made it successful, and we believed the loyalty it inspired gave us added value. We hadn't yet realized in the early 1980s that there was going to be a skill shortage, but we clearly needed to upgrade our training. A lot of our competitors, especially in the semiconductor business, hired people, used their skills, terminated them when their skills were out of date, then hired new people with new skills. But we had plants where 60% to 70% of the workers were Service Club members.

We didn't want to break a model that had worked for 50 years, but we had some people who thought that if they made that ten-year mark they could mentally retire, and that was an attitude we had to fix. In the end, we had to let people know that "poor performance" included an unwillingness to change. We had to abandon paternalism for shared responsibility.

A second major misconception was that senior managers needed only a briefing to understand the new quality systems. Conceptually, in fact, they grasped them very quickly and believed in their importance. But their behavior patterns didn't change, and that made life very difficult for middle managers.

Operations review is a good example. If a production team had mastered the new techniques and was eager to apply them, and if senior management paid lip service to quality but still placed its highest priority on shipping goals, middle managers got caught in the squeeze. Workers expected them to emphasize quality even if that made some deliveries late. Top management expected them to improve quality but not at the expense of schedule.

Workers began to wonder why they'd taken the training. They'd learned how to keep a Pareto chart and make an Ishikawa diagram, but no one ever appeared on the floor and asked to see one. On the contrary, some of their immediate managers wanted product shipped even if it wasn't perfect. Top managers, on the other hand, began to wonder why it was that people took the courses so carefully designed for them and then went back to their jobs and did nothing different. Shipping goals were being met, but quality was not improving.

At about this point in our frustration, we asked two universities to evaluate our return on investment. They identified three groups:

> In those few plants where the work force absorbed the whole curriculum of quality tools and process skills and where senior managers reinforced the training by means of new questions appropriate to the new methods, we were getting a $33 return for every dollar spent, including the cost of wages paid while people sat in class.
>
> Plants that made use of either the quality tools or the process skills but not both, and then reinforced what they taught, broke even.
>
> Finally, plants that taught all or part of the curriculum but failed to reinforce with follow-up meetings and a new, genuine emphasis on quality had a negative return on investment.

We were learning our first lesson all over again, that change is not just driven from the top—change must *begin* at the top. We had also begun to understand that one secret of manufacturing success was a language common to every employee, in this case a common language of quality. But if quality was to be the new language, every top manager had better learn to speak it like a native. Since Bob Galvin believed quality training was useless unless top managers gave quality even more attention than they gave quarterly results, he dramatized the point at operations review meetings. He insisted that quality reports come first, not last, on the agenda, and then he left before the financial results were discussed.

Thin Ice

By 1984, we were admittedly a little disappointed by what we had achieved so far, but we believed we could address the problem with a new, double-edged initiative. First, by hiring and training more carefully, we would bring our manufacturing talent up to the standard in product design, and second, having now learned our lesson about top executive involvement, we would offer the training to upper management as well as line workers.

There was no question that manufacturing was a second-class citizen. Our recruiting strategy had always been to find the best engineers, and the best engineers went into product design. We had never deliberately looked for the best manufacturing or materials-management people. We now consciously upgraded the status, rewards, and recruiting for Motorola manufacturing.

We also put together a two-week program of courses that tried to project what our kind of manufacturing would look like in ten years and to think about the changes we would have to make to stay competitive. We wanted all the decision makers in manufacturing to take part.

We began, as so often before and since, by repeating a past mistake. We assumed the proper people at the top would sign up without prompting. What we found was that many simply delegated the program to subordinates. Our advisory board—Galvin and 11 senior managers—soon saw what was happening and changed the policy. From then on, the board "invited" people to take the training, and board members began by inviting themselves, then everyone else at the top.

The curriculum centered on manufacturing technologies—computer-aided design, new quality measurements, computer-integrated manufacturing, just-in-time—as well as on what we called "unity of purpose," our name for the empowered work force. We also began to focus on time. We used to take three to seven years to design a new product; now we were shooting for 18 months. Finally, we started talking about a contract book where marketing, product design, and manufacturing would meet, argue, and reach genuine agreement about the needs of the market, the right new product, and the schedules and responsibilities of each group in producing it.

Historically, such agreements had often been made but never kept. The three departments would meet and agree, product design would go home and invent something quite different but wonderfully state-of-the-art, manufacturing would do a complete redesign to permit actual production in the real world, and finally marketing would find itself with a product bearing little resemblance to what it had asked for and been promised.

We wanted to use the training to send a message to the company about achieving quality through the integration of efforts across functions, a message not just about quality of product but about quality of people, quality of service, quality of the total organization.

In 1985, we also started an annual training and education event for senior management. Each year, the CEO picks a topic of critical concern and brings his top executives together to thrash it out with the help of some carefully chosen experts.

That first year the topic was competition, especially Asian competition. There were Asian companies—whole Asian countries—going after every market segment we were in. Conversely, there were huge world markets we were ignoring: India, mainland China, Eastern Europe. The goal was to scare the wits out of all the participants and wake them up to the perils and opportunities of our position. We had people calling themselves worldwide product or marketing managers who didn't even have a passport.

From 1985 to 1987, our top 200 people spent 17 days each in the classroom. Ten days on manufacturing, five days on global competition, and two more days on cycle-time management. We then drove that training down through the organization via the major components of the curriculum. What was ten days for a vice president might have been eight hours for a production worker, but eventually they were all speaking the same language. Or so we believed.

In 1985, we decided, after considerable fact finding and lengthy deliberation, to open our new cellular manufacturing facility in the

United States rather than take it offshore. Top management had just gone through its workshop on foreign competition, and there was a general feeling that, with our manufacturing expertise, we should be able to compete with anyone in the world. We acknowledged the fact that it wasn't only low-cost labor we'd been going offshore to get but brains—first-rate manufacturing know-how. In our naive self-satisfaction over the success of our training programs, we figured we now had across-the-board know-how in Illinois.

Our work force in Arlington Heights, outside Chicago, knew radio technology, and given the similarities, we believed these workers could make the bridge to cellular. In addition, they had improved quality tenfold in the first five years of training and were well on their way to doing it again. Nevertheless, we were making what was fundamentally an emotional commitment, not a hard-nosed business decision. However much quality had improved, it had probably not improved enough to compete in new technologies when our labor was going to cost us more than it cost our Asian competitors. However thoroughly the Arlington Heights workers understood radios, we were about to empower them to do much more than they had done in the past—not just one or two assembly functions but quality control, flexible manufacturing, and mentoring for the several thousand new hires we'd eventually have to add.

Maybe we sensed the thin ice we were on. In any case, we did a quick math assessment to see exactly where we stood with regard to further training. The scores were a shock. Only 40% passed a test containing some questions as simple as "Ten is what percent of 100?"

The Work Force That Couldn't Read

Let me dwell for a moment on the full drama of those results. The Arlington Heights work force was going to lead the company into global competition in a new technology, and 60% seemed to have trouble with simple arithmetic. We needed a work force capable of operating and maintaining sophisticated new equipment and facilities to a zero-defect standard, and most of them could not calculate decimals, fractions, and percents.

It took us several months and a number of math classes to discover that the real cause of much of this poor math performance was an inability to read or, in the case of many immigrants, to comprehend English as a second language. Those who'd missed the simple percentage question had been unable to read the words. In a sense, this was good news: it meant their math might not be as bad as we'd feared. But the news was bad enough even with the silver lining. For years, we'd been moving computers into the work line to let everyone interface with a keyboard and a screen; since 1980, the number of computer terminals at Motorola had gone from 5,000 to 55,000. At the new plant, workers would need to feed data into these terminals and extract information from them. Now we had reason to wonder if they'd be able to use their computers effectively.

And yet these people were superior employees who had improved quality tenfold and more. How had they done that if they couldn't read? It was a mystery—and a great problem.

The mystery was easily solved. In the early 1980s, we had had several layers of middle managers who acted as translators. They took the directions on the screens and put them into spoken English or, in many cases, Polish or Spanish, and the workers—dedicated, motivated people—carried them out.

The problem, however, remained. In fact, it grew rapidly more serious as we began to set up the new plant. We had begun to remove layers of middle management in the mid-1980s, and in setting up the new factory we wanted to leave no more than two or three levels between the plant manager and the greenest entry-level hire. For that to work, we had to have people with basic skills who were quick to learn—and quick to teach. After all, we couldn't staff the whole plant with our current employees, even if they all had the necessary skills. We had to have a cadre of 400 to 500 people who could serve as teachers and role models for others as we brought them in.

In the old days, our selection criteria had been simply, "Are you willing to work? Do you have a good record of showing up for work? Are you motivated to work?" We didn't ask people if they could read. We didn't ask them to do arithmetic. We didn't ask them to demonstrate an ability to solve problems or work on a team or do anything except show up and be productive for so many hours a week.

The obvious, drastic answer would have been to fire that old work force and hire people who could meet our standards, but that was not an acceptable approach for our company and probably not a realistic one given the educational level of the available labor pool. We quickly implemented a standard for new hires based on seventh grade math and reading skills, but as we recruited people from the outside, we found remarkably few who met that seventh grade standard.

We look for many things in our applicants, beyond literacy. For example, they must also be punctual and responsible. But at some installations, the requirements are even higher. In one location, we opened a sophisticated plant and found we had to screen 47 applicants to find 1 who met a ninth grade minimum requirement and also passed a drug test. This was admittedly an extreme situation, and the results were also skewed by the drug test. Still, at 1 applicant in 47, a company soon runs out of potential hires, especially if anyone else is competing for the labor supply.

We quickly saw that a higher set of standards for new hires was not the whole answer to our Illinois problem, particularly since we weren't willing to lay off dedicated people who'd been with us for years. And Arlington Heights was no isolated pocket. Documenting installations one by one, we concluded that about half of our 25,000 manufacturing and support people in the United States failed to meet the seventh grade yardstick in English and math.

At a plant in Florida, we offered English as a second language, thinking maybe 60 people would sign up, and we got 600—one out of three employees. By signing up, they were telling us they couldn't read the memos we sent them, the work order changes, the labels on boxes.

Joining Forces with the Schools

When we started MTEC in 1980, it was on the assumption that our people had the basic skills for the jobs they were doing. For five years, we saw ourselves as an agent of change, a school for the language of quality, a counselor and goad to senior management, a vocational instructor in the new skills and measurement systems that would enable our employees to lead the electronics industry in sophistication and quality control.

Our charter as an educational program was to provide continuous training to every employee. Then suddenly, to meet our business needs, we found we had to add remedial elementary education. We had never wanted to be in the grade school business, and it threw our investment strategy into chaos. An annual budget projected at $35 million to $40 million for the early 1980s had grown to $50 million a year by 1988, and we were going to need $35 million more over a three- to five-year period to correct the literacy problem.

Yet budget was only one of our problems. Morale was an even thornier issue. Remedial education made many people uncomfortable, and some were literally afraid of school, an environment in which they had failed as children. We didn't want to embarrass them; these weren't marginal workers, remember, but valuable, experienced employees. Yet they were also people with serious math and literacy problems, and we had to educate them.

We took the position that remedial math and language instruction was simply another form of skills training. We videotaped older employees within a few years of retirement and had them say to others their own age and younger, "Look, we're not going to get out of here without learning new skills. We need to go to class. Back in 1955, we had another moment of truth when we had to move from tubes to transistors. We did it then; we can do it now."

We also had Bob Galvin respond personally to letters. If a man wrote to ask if he really had to go back to school at 58 and study math, Bob would write back, "Yes, you do. But everyone does. You do your job well, but without more math you won't survive in the work environment another seven years."

We adopted a policy that said everyone had a right to retraining when technology changed. But if people refused the retraining, then we said we'd dismiss them. In fact, we had refusals from 18 employees with long service, and we dismissed all but one. That sent another strong message.

There was also a flip side to the policy. It said that if people took retraining and failed, then it was our job to figure out a way of helping them succeed. After all, nearly one-fifth of the U.S. population has some kind of learning disability. If some of our employees couldn't learn to read and compute in conventional classrooms, we tried to help them learn some other way. If they couldn't learn at all but had met our hiring standards 20 or 30 years ago and worked hard ever since, we found jobs for them at Motorola anyway.

Uncovering widespread math and reading problems was a great shock to the system, but it had the beneficial effect of pushing us into three watershed decisions that shaped the whole future of MTEC and Motorola. First, we realized that remedial elementary education was not something we could do well ourselves, so we turned for help to community colleges and other local institutions. Second, we decided to take a harder look at certain other skills. We had always assumed, for example, that high school and junior college graduates came to us equipped with technical and business skills like accounting, computer operation, statistics, and basic electronics. When we discovered they did not—by this time nothing surprised us—we decided to go back to the community colleges that were already helping us with remedial math and English. We told them what we needed, and they had courses with the corresponding titles, so we sent them our people.

Our next surprise was that the courses weren't quite what the titles implied. The community colleges had fallen behind. Their theories, labs, and techniques were simply not up to modern industrial standards. They hadn't known where we were going, and we hadn't bothered to tell them.

This discovery forced our third major decision—to begin building the educational partnerships and dialogues that eventually led us to Motorola University.

We've deliberately tried to treat our educational suppliers the same way we treat our component and chemical suppliers. To begin with, that means acknowledging that in a certain sense we buy what they produce. To make that work, they have to know what we need, and we have to know what not to duplicate. We exchange faculty, jointly develop curriculum, share lab equipment, and tend our mutual feedback mechanisms. In half our plants, as an experiment, we have vocational training experts—community college staff members whom we pay full-time—to act as the ongoing bridge between the changes going on in our business and the changes necessary in the college curriculum. We're also developing dialogues with engineering schools, as well as elementary and secondary schools, about what we need and how it differs from what they provide.

Dialogue means not a single meeting, or two or three, but regular meetings every few weeks. We get presidents, deans, superintendents, principals, professors, and teachers to sit down with our vice presidents for manufacturing and quality, our CFO, the managers of various plants and functions and talk about our needs and theirs.

Schools and colleges have no self-evident need to collaborate with us to the extent we'd like them to. Some educators and academics believe businesspeople are universally unprincipled, that the reason for our involvement in education is to serve ourselves at the expense, somehow, of the community at large.

Yet in the course of talking and interacting, we've found that the benefits to all parties have emerged quite clearly. Some of our strongest support comes from the people who teach remedial English and math. They know how widespread the problem is beyond Motorola and seem grateful that we're eager to attack illiteracy without assessing blame. The point is not that our concern with education is utterly altruistic but that better and more relevant education helps business, labor, and the schools themselves.

This last point is important. Take the case of a community college. To begin with, we supply them with, say, a thousand students a year, and a thousand tuitions. We also donate equipment. More important than either of these benefits, perhaps, is the fact that we offer an insight into what the globally competitive marketplace demands of students and educators. Most colleges are very resource poor. But the dialogue means that they can come into our plants, talk to us, study state-of-the-art manufacturing, and they can use our facilities to pursue their own staff development. We provide summer internships for their teachers and set aside a certain number of in-house training slots each year for community college faculty. We encourage them to use our labs and equipment to teach not only Motorolans but all their students. We believe—and I think they believe—that the mutual benefits are huge.

We've assigned one of our senior people to be the director of institutional relationships. His job is to understand and try to improve the supply lines that run from elementary schools, high schools, and colleges to Motorola. He's there to diagnose their needs, communicate ours, and fund projects that bridge the gap. Motorola has set up an account—in addition to the $60 million education budget—that we can use to invest in schools that are willing to undertake change and to work with us to address the needs of the populations that supply our work force.

For example, two of our scientists came up with an electronics kit that costs only $10 to produce. Then we discovered that the average high school physics teacher has only $2 per student to spend. So we decided to use the account to donate the kits to schools we draw students from and to teach the teachers how to use them.

We also sponsor a planning institute for superintendents from 52 school districts in 18 states on how to build strategic education plans and work with their business constituencies, of which we are one element. We're also trying to build a volunteer corps of Motorola employees who want to work with schools and teach, plan, or translate particular classes into industry-specific programs.

Open-Ended Education

In 1980, we thought we could provide x amount of training, get everybody up to speed, then back off. Now we know it's open-ended. Whenever we reach a certain level of expertise or performance, there's always another one to go for. Now we know it's a continuous investment from both sides—from the individuals who attend classes and apply new skills and from the corporation that designs new training programs and makes the time available to take them. In fact, we now know there is no real distinction between corporate education and every other kind. Education is a strenuous, universal, unending human activity that neither business nor society can live without. That insight was another that led us to Motorola University.

When Bob Galvin first suggested a university in 1979, the company wasn't ready. Galvin started interviewing presidents of real universities for advice and leads, but in the meantime I interviewed 22 senior Motorola executives to get their reactions. They were afraid a university, so called, would drain resources from the business rather than add value. For my own part, I was afraid the name "university" was too pretentious. This wasn't to be a seat of free and open inquiry. This was to be training and education for work force and managers.

Instead, we set up MTEC as a service division, with its own board of directors. That made us the only corporation in the United States with a training advisory board that included the CEO and other senior officers. But it didn't make us a university.

Then in 1989, CEO George Fisher made the suggestion. He believed the word university would give us greater autonomy. He also thought it would create an expectation we'd have to grow into.

A great deal had changed in those nine years. First of all, most of our senior clientele—the 22 executives and their fellow skeptics—had stopped seeing education as a cost and had begun to accept it as an indispensable investment. They'd seen returns. Most other people in the company had seen more than that. They'd seen themselves picking up marketable skills; they'd felt themselves growing in self-esteem and self-confidence.

Another difference was that by 1989 we were working closely with the schools, and when public educators heard the word university, their response was positive, even enthusiastic. They took it to mean that we were serious about education, not just enthralled with our own bottom line.

But what *was* a university? And more to the point, what was a Motorola University? What kind of model should we work from? There were several that might apply, it seemed to me.

One came from the charter of the City University of New York, which says that one of the university's central missions is to meet the needs of the city's residents. What we had was a training department that focused on the needs of the organization. Becoming a university would mean a shift toward meeting the needs of individuals, the "residents" of the corporation.

Another possible model was the open university of the United Kingdom, which, instead of bringing people to the education, takes the education to the people and puts it in a context they can understand.

A third possibility was the model described by Cardinal Newman in *The Idea of a University*, which, after 150 years, is still the cornerstone of liberal education. Newman's ideal university had no place for vocational training, so in that sense he and we part ways. But in another sense, we're in complete agreement. Newman wanted his university to mold the kind of individual who can "fill any post with credit" and "master any subject with facility"—an excellent description of what we wanted Motorola University to do.

Our evolving vision of the university contains elements of all three models. We try to make our education relevant to the corporation, to the job, and to the individual. We also try to bridge the gap between our company and the institutions that supply us with people. We don't intend to grant degrees, but we do intend to design courses that accrediting boards will certify and that the universities that give degrees will count. (See "Curriculum Development at Motorola U.")

Curriculum Development at Motorola U

Motorola never farms out curriculum design—the road map that shows what we need to learn. But we may contract with an individual, a group, or a college to design a particular course, which we then turn over to our deans of delivery—one for the United States, one for the rest of the world—for distribution. Those deans search out instructors through the colleges or just hire free agents, whom we train. Sometimes, too, courses are packaged as software, as videos, or go out via satellite.

For each functional area we have a course-design team of instructional system designers whom we think of as product engineers in terms of title, pay, status, and job security. Their expertise is knowing how to extract the skill requirements from a job analysis and package them into training for adults.

Of 1,200 people involved in training and education at Motorola—including 110 full-time and 300 part-time staff at Motorola University—23 are product design engineers. They're like department chairs, and the senior product manager is like a dean. The faculty are the actual course writers, audiovisual specialists, and instructors.

We divide each functional curriculum—engineering, manufacturing, sales and marketing—into three parts: relational skills, technical skills, and business skills. M.U. itself is responsible for teaching relational skills—customer satisfaction, effective meetings, effective manufacturing supervision, negotiation, effective presentations. Since 1985, however, we have developed the curriculum for technical and business skills—basic math, electronics, accounting, computer operation, statistical process control—in cooperation with community colleges and technical schools.

As a result of what we've learned over the last decade, the Motorola curriculum has two more critically important elements. The first is culture. Culture in a corporation is partly a question of history and of common language—a form of tribal storytelling. At Motorola, for example, we believe in taking risks. Our history shows it was one of the things that made us great. Telling those stories creates a tradition of risk, invests it with value, and encourages young people to go out and do the same.

But culture is not only inherited, it's also created. Teamwork is an example. Our history is one of individual contributors, not teams. So training has to emphasize team building and downplay the Lone Ranger culture we valued in the 1950s and 1960s.

The second extra element is remedial reading and math, which today means algebra. In most of our factories, we want analyses and experiments requiring algebra done right on the line. As for reading, we've found that the engineering group writes work order changes, product specifications, and product manuals at a fourteenth to seventeenth grade level— to be read by people who read at the fourth to sixth grade level. We're trying to lower the former and raise the latter. When we talk about a literate work force by 1992, we mean seventh grade. But our goal remains ninth grade by 1995.

We are entering a new era of partnership with established universities. We give them feedback on the courses they teach and even on their faculties. Most companies have just blindly reimbursed tuition costs, but we think we have a legitimate interest in the schools that prepare our future employees and in the colleges that provide continuing adult education at our expense.

For example, Motorola and Northwestern jointly designed a quality course for the second year of Northwestern's MBA program. Northwestern also offers courses taught half-term by one of its professors and half-term by one of our experts—first half, theory; second half, application. We work closely with the Illinois Institute of Technology as well, and with Arizona State University. For the first time, we actually evaluate university services, not just assume they must be good.

We avoid collegiate trappings—no professors or basketball teams—but we do make academic appointments of a kind: we name people to staff positions, and for two or three years they leave their jobs and devote themselves to educational activities. (See "Finding and Training Faculty.") We also have a Motorola University Press that prints more than a million pages a month and plans to publish a series of books on design and quality written by one of our employees.

Finding and Training Faculty

I own a videotape of an all-day engineering class whose instructor, in closing the session, thanks the students for their active participation. In fact, except to breathe, not one of them has made a sound. In eight hours of teaching, the instructor never asks a question. He never makes eye contact. He is on top of his subject technically, but the class is boring beyond belief.

The teachers at Motorola University aren't there to implant data. They're there to transfer information and get it applied quickly. We design curricula and train teachers with that end in mind.

Training and certification of all teachers is in four phases of 40 hours each. The first phase is procedural: how to use flip charts, foils, chalkboards; how to manage a classroom; how to ask questions; how to observe, listen, and reinforce.

The second phase deals with the particular kinds of behavior we want to see teachers and students exhibit. We believe in participative management, so instructors must be models of openness and respect for other opinions. We also emphasize teamwork, so teachers must encourage mutual support and try to curb the competitive, who's-best model most people learned in school.

The third 40 hours deals with subject matter. We walk teachers through the course and then we walk them through the research that underlies it. We don't want them to teach *their* version of, say, Effective Meetings; we want them to teach *ours*. Not everyone can deliver on those terms. For example, few academics can do it our way. They're used to interpreting material independently, so after the first page, it tends to take on their own particular slant. It may make a fascinating course, but we can't have 3,000 people learning 35 different versions of Effective Meetings.

Finally, we assign teachers to master instructors who use them as coteachers and then have them teach on their own with feedback before turning them loose.

We have found three groups of people who make especially good teachers. The first is recently retired Motorola employees. We start talking to them six months out and try to get them through certification before retirement. Those who make it are superb. They believe in what they're teaching, they know how to apply it, and they can tell the stories from the past and relate them to the future.

We have one powerful, middle-management course on managing change—specifically, the change that must take place if Motorola is to prosper over the next decade—that is taught by insiders only. Bill Weisz, a former CEO and now a vice chairman of the board, devotes the last seven hours to the topic "I built this bureaucracy, and if I was 25 years old, here's how I'd blow it up and build what we need for the next ten years."

The second group consists of married women with college degrees whose children have left home. Lack of experience makes it hard for them to find jobs, so we've worked out programs with community colleges to screen them and reorient them to the business world. In courses like Effective Meetings, Interpersonal Skills, and Basic Sales Skills, they're our best performers by far. Most students think they're old business hands, but their actual experience—community politics, selling ideas to children, managing teenagers—is just as rich and certainly just as useful.

The third group represents a kind of talent sharing. Like most other companies, Motorola has a rule against rehiring people who've taken early retirement. Now we trade early retirement talent with other companies. AT&T gives us a list of 100 potential instructors who've just left, and we give them our list. It's been a success. People like translating their experience from AT&T or IBM and making it relevant and effective in our different world. Eventually we hope to change the rules so that people who pass our certification will always have an alternative to early retirement.

As for Newman's grander concept of the university, our commitment is not to buildings or a bureaucracy but to creating an environment for learning, a continuing openness to new ideas. We do teach vocational subjects, but we also teach supervocational subjects—functional skills raised to a higher level. We not only teach people how to respond quickly to new technologies, we try to commit them to the goal of anticipating new technologies. We not only teach people how to lead a department to better performance and higher quality, we try to dedicate them to the idea of continuing, innovative leadership in the workplace and the marketplace. We not only teach skills, we try to breathe the very spirit of creativity and flexibility into manufacturing and management.

Electrifying the Now and Future Employee

The word university is undeniably ambitious, but Motorola management has always tried to use words in ways that force people to rethink their assumptions. The term university will arouse curiosity and, I hope, raise the expectations of our work force and our training and education staff. We could have called it an educational resource facility, but who would that have electrified?

As a first step, we have decided that Motorola University will be a global institution. We are already working on a formal relationship with the Asia Pacific International University, based in Macao. Motorola University is presently open to the employees of our suppliers, of our principal customers, and even of our educational partners, but we envision a time when the university will accept students from outside our immediate community of companies and institutions, people who will not necessarily work for Motorola or any of our suppliers or customers at any time in their lives.

At the same time, one of our goals is to have the best graduates of the best institutions want to work for us. To achieve that, we must attract young people into our classrooms so they'll know how good we are, and, just as important, so they'll know they have to take certain steps in their own development in order to work for us or someone like us as adults.

One of the things we have on the drawing board is a math-and-science summer institute for fifth, sixth, and seventh graders. Another is a simulation using an aircraft from the movie *Top Gun*. We'll stop the plane in mid-takeoff from a carrier and dissect it, showing the geometric and algebraic principles used in its design, then relate that back to the computer to show the same principles used in the design of other things. The goal is not to teach kids design but to show them that algebra is worth suffering through.

The point of a program like that is not to get kids to work for Motorola. We'd like the really inspired ones to want to work for us when that time comes. But if not, well, at least they'll be able to work for someone.

3
Time-and-Motion Regained

Paul S. Adler

Standardization is the death of creativity.

Time-and-motion regimentation prevents continuous improvement.

Hierarchy suffocates learning.

U.S. manufacturing is in the throes of revolution, and assumptions like these are becoming the new conventional wisdom about work. This new gospel sets up Frederick Winslow Taylor and his time-and-motion studies as the villain. It asserts that quality, productivity, and learning depend on management's ability to free workers from the coercive constraints of bureaucracy. It insists that detailed standards, implemented with great discipline in a hierarchical organization, will inevitably alienate employees, poison labor relations, stifle initiative and innovation, and hobble an organization's capacity to change and to learn.

But what if, as I believe, this new creed is wrong? What if bureaucracy can actually be designed to encourage innovation and commitment? What if standardization, properly understood and practiced, should prove itself a wellspring of continuous learning and motivation?

In Fremont, California, a GM-Toyota joint venture called New United Motor Manufacturing Inc., NUMMI, for short, has succeeded in employing an innovative form of Taylor's time-and-motion regimentation on the factory floor not only to create world-class productivity and quality but also to increase worker motivation and satisfaction. What's more, NUMMI's intensely Taylorist procedures appear to encourage rather than discourage organizational learning and, therefore, continuous improvement.

This outcome seems surprising because for decades our attitudes toward work have been shaped by a chain of reasoning that has led us to expect (and guaranteed that we would get) a vicious circle of escalating managerial coercion and employee recalcitrance. The reasoning runs something like this:

> When tasks are routine and repetitive, efficiency and quality require standardized work procedures.
>
> High levels of standardization rob jobs of their intrinsic interest, reducing motivation and creativity.
>
> Demotivating work leads to dysfunctional employee behavior such as absenteeism, high turnover, poor attention to quality, strikes, even sabotage.
>
> Counterproductive behavior by the work force requires more authoritarian management, more hierarchical layers, and even higher levels of standardization.

In short, Taylorism leads inevitably to work-force discontent and union belligerence, which in turn lead inevitably to higher levels of bureaucratic excess. The organization of work comes to build on the dehumanizing logic of coercion and reluctant compliance. Meanwhile, quality, profits, and job satisfaction all suffer.

NUMMI's experience flies directly in the face of this thinking. That's because the second step in this chain of reasoning is false. Formal work standards developed by industrial engineers and imposed on workers are alienating. But procedures that are designed by the workers themselves in a continuous, successful effort to improve productivity, quality, skills, and understanding can humanize even the most disciplined forms of bureaucracy. Moreover, NUMMI shows that hierarchy can provide support and expertise instead of a mere command structure.

What the NUMMI experiment shows is that hierarchy and standardization, with all their known advantages for efficiency, need not build on the logic of coercion. They can build instead on the logic of learning, a logic that motivates workers and taps their potential contribution to continuous improvement.

In practice, NUMMI's "learning bureaucracy" achieves three ends. First, it serves management by improving overall quality and productivity. Second, it serves workers by involving them in the design and control of their own work, increasing their motivation and job satisfaction, and altering the balance of power between labor and management. Third, it serves the interests of the entire organization—management and the work force—by creating a formal system to encourage learning, to capture and communicate innovation, and to institutionalize continuous improvement.

The Worst Plant in the World

NUMMI is housed in what was once the General Motors assembly plant in Fremont, California, 35 miles southeast of San Francisco, which opened in 1963 and manufactured GM trucks and the Chevy Malibu and Century. At the old GM-Fremont plant, work was organized along traditional Taylorist lines, with more than 80 industrial engineers establishing assembly-line norms that management then did its best to impose on the work force, with the predictable results.

Over the years, GM-Fremont came to be what one manager called "the worst plant in the world." Productivity was among the lowest of any GM plant, quality was abysmal, and drug and alcohol abuse were rampant both on and off the job. Absenteeism was so high that the plant employed 20% more workers than it needed just to ensure an adequate labor force on any given day. The United Auto Workers local earned a national reputation for militancy; from 1963 to 1982, wildcat strikes and sickouts closed the plant four times. The backlog of unresolved grievances often exceeded 5,000.

GM-Fremont reached its peak employment of 6,800 hourly workers in 1978. Numbers then declined steadily to a little over 3,000 when GM finally closed the plant in February 1982.

Discussions between GM and Toyota about a possible joint venture began that same year. In February 1983, the two companies reached an agreement in principle to produce a version of the Toyota Corolla, renamed the Nova, at the Fremont plant, using Toyota's production system. GM would be responsible for marketing and sales; Toyota would take on product design, engineering, and daily operations. The new entity, NUMMI, would manufacture and assemble the car. Beginning in 1986, the plant also made Corolla FXs. In 1988, both the Nova and the FX were phased out, and Fremont began building Corollas, Geo Prizms, and, as of late 1991, Toyota trucks.

The two companies' objectives were complementary. GM wanted to learn about Toyota's production system. It also obtained a high-quality subcompact for its Chevrolet division at a time when GM's market share was rapidly eroding. Toyota wanted to help defuse the trade issue by building cars in the United States. To do this, it needed to learn about U.S. suppliers.

Toyota later claimed it had also wanted "to gain experience with American union labor," but at first Toyota wanted nothing to do with the UAW. As it happened, there was no alternative. GM offered them no other facility, and the UAW had de facto control of Fremont. Moreover, GM was afraid of a union backlash at other plants if it tried to set up the joint venture as a nonunion shop.

In September 1983, NUMMI and the union signed a letter of intent recognizing the UAW as sole bargaining agent for the NUMMI labor force, specifying prevailing auto-industry wages and benefits, and stipulating that a majority of the work force would be hired from among the workers laid off from GM-Fremont. In return, the UAW agreed to support the implementation of a new production system and to negotiate a new contract.

NUMMI was formally organized in February 1984. Toyota contributed $100 million in capital, and GM supplied the Fremont plant. Hiring began in May. Every applicant went through three days of production simulations, written examinations, discussions, and interviews. Managers and union officials jointly evaluated applicants for the hourly jobs: team leader and team member. The union also played a role in selecting managers, except for the 16 who came directly from GM and a group of about 30 Toyota managers and production coordinators who came from Japan. The CEO, Tatsuo Toyoda, brought with him the prestige of the company's founding family.

Over the following 20 months, NUMMI hired 2,200 hourly workers—85% from the old GM-Fremont plant, among them the old union hierarchy. (Almost none of GM-Fremont's salaried employees was rehired. In any case, many had long since moved to other GM plants.) Since GM-Fremont had done little hiring for several years before it closed, the average age of the new work force was 41. Most had high school educations. About 26% were Hispanic, 20% black, and 15% female.

The first group of 450 team leaders and the entire NUMMI management team attended a three-week training program at the Toyota plant in Japan—Takaoka—on which NUMMI was modeled. These people then helped to set up the new plant and train workers.

The NUMMI production system required people to work harder than they had at GM-Fremont. Jobs at the old plant occupied an experienced worker about 45 seconds out of 60. NUMMI's norm is closer to 57 seconds out of 60. And because workers have to meet much higher quality and efficiency standards, they have to work not only harder but smarter as well.

By the end of 1986, NUMMI's productivity was higher than that of any other GM facility and more than twice that of its predecessor, GM-Fremont. In fact, NUMMI's productivity was nearly as high as Takaoka's, even though its workers were, on average, ten years older and much less experienced with the Toyota production system. Quality, as rated by internal GM audits, customer surveys, and *Consumer Reports* was much higher than at any other GM plant and, again, almost as high as Takaoka's.

Equally important, absenteeism has dropped from between 20% and 25% at the old GM-Fremont plant to a steady 3% to 4% at NUMMI; substance abuse is a minimal problem; and participation in the suggestion program has risen steadily from 26% in 1986 to 92% in 1991. When GM-Fremont closed its doors, it had more than 2,000 grievances outstanding. As of the end of 1991, some 700 grievances had been filed at NUMMI altogether over the course of eight years. The overall proportion of employees describing themselves as "satisfied" or "very satisfied" has risen progressively to more than 90%.

In 1990, Toyota announced that it would invest $350 million in an additional assembly line to build a Toyota truck for the U.S. market. So NUMMI hired 650 hourly workers on top of the 3,100—plus 400 salaried personnel—already employed. The first trucks rolled off the line in August 1991.

Fear, Selection, Socialization

NUMMI's remarkable turnaround poses an obvious question: How is it possible to convert a plant from worst to best quality and from dismal to superlative productivity over the course of a few months? The most obvious answers are not entirely satisfying.

For example, fear. The GM-Fremont plant closed in 1982, and the people rehired by NUMMI didn't go back to work until 1984. Two years of unemployment can produce a great deal of cooperation. In fact, some NUMMI workers believe management makes deliberate use of the specter of another plant closure as a veiled threat to keep people in line. But the chairman of the union bargaining committee points out that while the old plant's closure obviously made workers more receptive to NUMMI's new approach, a return to old coercive management methods would have produced a rapid return to old antagonistic work-force behavior patterns.

A second possibility is that management weeded out troublemakers in the rehiring process. But in fact NUMMI rehired the entire union hierarchy and many well-known militants. In general, very few applicants were screened out. The union even won a second chance for some who failed drug tests the first time around.

A third answer is that NUMMI made use of a comprehensive socialization process during hiring to instill a new set of values in the new work force. Certainly, NUMMI did its best to shape and alter the attitudes of both workers and managers. For example, the company tried to undercut the customary we-they divisions between workers and management by eliminating special parking and eating facilities for managers and by introducing an identical dress code—uniforms—for everyone. Management also devoted a great deal of attention to each individual hire and welcomed each personally to the company that was going to build "the finest vehicles in America."

However much these three factors—fear of unemployment, selection, and socialization—may have contributed to the final outcome, they do not adequately explain NUMMI's continuing success or its ability to let workers draw improved motivation and greater satisfaction from a system that places them in a more regimented and bureaucratic environment and makes them work harder and faster. The most critical piece of that explanation lies in the production system itself and in the policies and practices that buttress it.

The NUMMI Production System

The idea of a production *system* is itself something of a novelty in many U.S. manufacturing plants. All factories have production techniques, procedures, and policies, but these usually comprise not so much a system as an ad hoc accumulation of responses to changing and often contradictory business and design demands. NUMMI's production system is a finely tuned, superbly integrated whole, refined by Toyota over decades of manufacturing experience.

The basic techniques are familiar at least in name. The assembly line is a just-in-time operation that does away with work-in-progress and makes quality assurance the responsibility of each work station. The application of *kaizen*, or continuous improvement, includes an extraordinarily active suggestion program, constant refinement of procedures, and the designation of special kaizen teams to study individual suggestions or carry out specific improvement projects. Every machine and process is designed to detect malfunctions, missing parts, and improper assemblies automatically. Every job is carefully analyzed to achieve maximum efficiency and quality. Job rotation is standard; workers are cross-trained in all team assignments and then allowed to shift from one task to another. Planned production leveling eliminates variation in daily and weekly schedules.

This system is essentially the same one Toyota uses in Japan, the same one many American manufacturers are now beginning to adopt. But NUMMI's approach is distinctive in two respects: first, its strong commitment to the social context in which work is performed, and, second, its intense focus on standardized work.

In terms of social context, NUMMI seeks to build an atmosphere of trust and common purpose. NUMMI maintains exceptional consistency in its strategies and principles, it carefully builds consensus around important decisions, and it has programs ensuring adequate communication of results and other essential information.

The basic structural unit is the production team, of which NUMMI has approximately 350, each consisting of five to seven people and a leader. The idea is that small teams encourage participative decision making and team bonding. Four teams comprise a group, led by a group leader who represents the first layer of management.

Above and beyond the production teams, the bigger team is everyone—all the workers, team leaders, managers, engineers, and staff in the plant as well as NUMMI's suppliers. Toyota leadership wants workers to understand that the company is not the property of management but of everyone together. In NUMMI's view, the primary purpose and responsibility of the management hierarchy is to support the production teams with problem-solving expertise.

The most substantive expression of this big-team strategy is the no-layoff policy spelled out in NUMMI's collective-bargaining agreement with the union. Recognizing that "job security is essential to an employee's well being," NUMMI agrees "that it will not lay off employees unless compelled to do so by severe economic conditions that threaten the long-term viability of the Company." NUMMI agrees to take such drastic measures as reducing management salaries and assigning previously subcontracted work to bargaining unit employees before resorting to layoffs.

Management sees the no-layoff policy as a critical support for its overall production strategy not only because it reinforces the team culture, but also because it eliminates workers' fear that they are jeopardizing jobs every time they come up with an idea to improve efficiency.

Workers came to trust this no-layoff commitment when in 1988 poor sales of the Nova brought capacity utilization down to around 60%. Workers no longer needed on the assembly line were not laid off but instead assigned to kaizen teams and sent to training classes.

Another important support for NUMMI's team concept is its radically simplified job classification system. Where GM-Fremont had 18 skilled trades classifications, NUMMI has two. Where GM-Fremont had 80 hourly pay rates, at NUMMI all production workers get the same hourly rate—currently $17.85—regardless of their jobs, except that team leaders get an extra 60 cents. There are no seniority-, performance-, or merit-based bonuses. Important as money is, equity is more important still in reducing tensions and resentments.

The second distinctive feature of NUMMI's system is standardization. Typically, American companies approach team empowerment by allowing teams considerable autonomy in how they accomplish tasks. NUMMI, in contrast, is obsessive about standardized work procedures. It sees what one NUMMI manager has called "the intelligent interpretation and application of Taylor's time-and-motion studies" as the principal key to its success. The reference to Taylor may be jarring, but it fits.

Standardized Work . . .

At GM-Fremont, industrial engineers did all time-and-motion analysis and formal job design, and workers tended to view them with resentment or contempt. The problem, as one union official described it, was that management assumed a "divine right" to design jobs however it saw fit. Industrial engineers with no direct experience of the work beyond capsule observation would shut themselves in a room, ponder various potentials of the human body, time the result, and promulgate a task design. Or so it seemed to workers, whom no one ever consulted despite their intimate familiarity with the specific difficulties of the work in question.

Normally, when an industrial engineer presented one of these pedantically designed jobs to a supervisor, the supervisor would politely accept it, then promptly discard it in favor of the more traditional kick-ass-and-take-names technique. The worker, in turn, usually ignored both engineer and foreman and did the job however he or she was able—except, of course, when one of them was looking. If an industrial engineer was actually "observing"—stop-watch and clipboard in hand—standard practice was to slow down and make the work look harder. The entire charade was part of an ongoing game of coercion and avoidance. Multiply this scenario by two shifts and thousands of workers, and the result is anything but the rational production of a high-quality car.

At NUMMI, in radical contrast to GM-Fremont, team members themselves hold the stopwatch. They learn the techniques of work analysis, description, and improvement. This change in the design and implementation of standardized work has far-reaching implications for worker motivation and self-esteem, for the balance of power between workers and management, and for the capacity of the company to innovate, learn, and remember.

The job design process itself is relatively simple. Team members begin by timing one another with stopwatches, looking for the safest, most efficient way to do each task at a sustainable pace. They pick the best performance, break it down into its fundamental parts, then explore ways of improving each element. The team then takes the resulting analyses, compares them with those of the other shift at the same work station, and writes the detailed specifications that become the standard work definition for everyone on both teams.

Taking part in the group's analytical and descriptive work involves every team member in a commitment to perform each task identically. In one sense, therefore, standardized work is simply a means of reducing variability in task performance, which may seem a relatively trivial achievement. In fact, however, reduced variability leads to a whole series of interconnected improvements:

Safety improves and injuries decline because workers get a chance to examine all the possible sources of strain and danger systematically.

Quality standards rise because workers have identified the most effective procedure for each job.

Inventory control grows easier, and inventory carrying costs go down because the process flows more smoothly.

Job rotation becomes much more efficient and equitable, which makes absences less troublesome.

Flexibility improves because all workers are now industrial engineers and can work in parallel to respond rapidly to changing demands. For example, NUMMI can convert to a new line speed in four to six weeks, a process that might easily have taken six months to a year at GM-Fremont, with its engineers frantically recalculating thousands of tasks and trying to force the new standards on workers. In fact, GM-Fremont never even attempted anything as demanding as a line-speed change. If orders declined, GM-Fremont had to lay off an entire shift. NUMMI's new capacity to alter line speed means, among other things, that the plant can accommodate a drop in orders by slowing production.

Standardized work also has the overall benefit of giving control of each job to the people who know it best. It empowers the work force. Not surprisingly, NUMMI discovered that workers bought into the process quite readily. As one manager put it, "They understood the technique because it had been done to them for years, and they liked the idea because now they had a chance to do it for themselves."

. . . and Continuous Improvement

Yet by far the most striking advantage of standardized work is that it gives continuous improvement a specific base to build on. As one manager put it, "You can't improve a process you don't understand." In this sense, standardization is the essential precondition for learning.

Indeed, standardization is not only a vehicle and a precondition for improvement but also a direct stimulus. Once workers have studied and refined their work procedures, problems with materials and equipment quickly rise to the surface. Moreover, since each worker is now an expert, each work station is now an inspection station—and a center of innovation.

At GM-Fremont, worker suggestions were apt to meet a brick wall of indifference. At NUMMI, engineers and managers are meant to function as a support system rather than an authority system. When a team can't solve a problem on its own, it can seek and get help. When a worker proposes complex innovation, engineers are available to help assess the suggestion and design its implementation.

The difference between traditional Taylorism and the learning-oriented NUMMI version resembles the difference between computer software designed to be "idiot-proof" and the kinds of computer systems that are meant to leverage and enhance their users' capabilities. The first "de-skills" the operator's task to an extent that virtually eliminates the possibility of error, but it also eliminates the operator's ability to respond to unpredictable events, to use the system in novel ways or adapt it to new applications. The idiot-proof system may be easy to use, but it is also static and boring. Leveraging systems make demands on the operator. They take time to learn and require thought and skill to use, but they are immensely flexible, responsive, and satisfying once mastered.

The difference goes deeper yet. At GM-Fremont—where work procedures were designed to be idiot-proof—the relationship between production system and worker was adversarial. Standards and hierarchy were there to coerce effort from reluctant workers. If the system functioned as expected and the operator was sufficiently tractable and unimaginative, the two together could turn out a fair product. There was little the operator could improve on, however, and the role of the system was utterly rigid until it broke down, whereupon everything stopped until a specialist arrived.

At NUMMI, the relationship of workers to the production system is cooperative and dynamic. Instead of circumventing user intelligence and initiative, the production system is designed to realize as much as possible of the latent collaborative potential between the workers and the system.

Suggestion programs illustrate the two approaches to organizational technology design. At many companies, suggestion programs are idiot-proof and opaque. They are designed primarily to screen out dumb ideas, and the basic review criteria, the identity of the judges, the status of proposals, and the reasons for rejection are all a black box as far as the workers are concerned. Predictably, a lot of these programs sputter along or die out altogether.

At NUMMI, the program is designed to encourage a growing flow of suggestions and to help workers see and understand criteria, evaluators, process, status, and results. Like a computer system designed to leverage rather than de-skill, the program helps employees form a mental model of the program's inner workings. Not surprisingly, workers made more than 10,000 suggestions in 1991, of which more than 80% were implemented.

In systems that de-skill and idiot-proof, technology controls, indeed dominates, workers. In systems designed for what experts call usability, the operator both learns from and "teaches" the technology. Using learned analytical tools, their own experience, and the expertise of leaders and engineers, workers create a consensual standard that they teach to the system by writing job descriptions. The system then teaches these standards back to workers, who, then, by further analysis, consultation, and consensus, make additional improvements. Continual reiteration of this disciplined process of analysis, stand-ardization, re-analysis, refinement, and restandardization creates an intensely structured system of continuous improvement. And the sa-lient characteristic of this bureaucracy is learning, not coercion.

This learning orientation captures the imagination. People no one had ever asked to solve a problem, workers who never finished high school, men and women who had spent 20 years or more in the auto industry without a single day of off-the-job training found themselves suddenly caught up in the statistical analysis of equipment downtime, putting together Pareto charts. One worker reported that he did liter-ally a hundred graphs before he got one right.

A woman on the safety committee in the body shop described how she applied kaizen techniques to her kitchen at home after a fire on her stove. She analyzed the kitchen layout, installed a fire extin-guisher, and relocated her pot tops so she could use them to smother flames. In short, she subjected herself and her home work space to the formal problem-solving procedures she had learned at the NUMMI plant.

The paradoxical feature such stories have in common is their enthu-siasm for a form of disciplined behavior that both theory and past practice seem to rule out. This paradox grows from our failure to distinguish between what Taylorist, bureaucratic production systems *can* be and what, regrettably, they have usually been.

The Psychology of Work

The chain of reasoning by which disciplined standardization leads inescapably to coercion, resentment, resistance, and further coercion seems to turn Taylorism and bureaucracy into what sociologist Max Weber called an iron cage. Taylorism and bureaucracy may have a devastating effect on innovation and motivation, the reasoning goes, but their technical efficiency and their power to enforce compliance seem to be the perfect tools for dealing with employees assumed to be recalcitrant. Taylor himself at least occasionally endorsed this coercive view of work. Italics bristling, he once wrote, "It is only through the *enforced* standardization of methods, *enforced* adoption of the best im-plements and working conditions, and *enforced* cooperation that this faster work can be assured. And the duty of enforcing the adoption of standards and of enforcing this cooperation rests with the *management* alone."

Against this background, it is hardly surprising that most managers and academics, at least in the West, have come to believe that Taylorism and bureaucracy will inevitably alienate workers and squander their human potential. But the psychological assumption underlying this expectation is that workers are incapable of delayed gratification. Managers seem to believe that performance will improve only as work comes more and more to resemble free play—our model of an intrinsically motivating activity. Indeed, it is an elementary axiom of economics that work is something that workers will always avoid.

NUMMI demonstrates the error of imputing infantile psychology to workers. Interviews with NUMMI team members suggest, in fact, that this whole historical accumulation of assumptions obscures three sources of adult motivation that the NUMMI production system successfully taps into:

First, the desire for excellence.

Second, a mature sense of realism.

Third, the positive response to respect and trust.

The first of these—the desire to do a good job, the instinct for workmanship—comes up again and again in conversations with workers. The NUMMI production system and the training that went with it increased both the real competence of workers and their feelings of competence. Workers talk a lot about expertise, pride, and self-esteem. One UAW official named "building a quality product" as one of the strategic goals that the union found most compelling at NUMMI. Perhaps the most striking story about pride in all the interviews came from a team leader (also see "Voices from the Factory Floor"):

> Before, when I saw a Chevy truck, I'd chuckle to myself and think, "You deserve that piece of crap if you were stupid enough to buy one." I was ashamed to say that I worked at the Fremont plant. But when I was down at the Monterey Aquarium a few weekends ago, I left my business card—the grunts even have business cards—on the windshield of a parked Nova with a note that said, "I helped build this one." I never felt pride in my job before.

The second element of motivation is a mature sense of realism—in this case, the understanding that unless NUMMI constantly improves its performance, competitors will take its market and its workers' jobs. A useful psychological theory cannot assume that workers are so captive to the pleasure principle that their only source of motivation is the immediate pleasure of intrinsically meaningful work. The evidence suggests that at least some of the workers at NUMMI are powerfully motivated by the simple recognition that international competition now forces them to "earn their money the old-fashioned way."

Voices from the Factory Floor: Excerpts from Interviews with Managers, Workers, and Union Officials

Team Leader

I'll never forget when I was first hired by GM many years ago. The personnel manager who hired us got the . . . workers who were starting that day into a room and explained: "You new employees have been hired in the same way we requisition sandpaper. We'll put you back on the street whenever you aren't needed any more." How in the hell can you expect to foster a loyal and productive work force when you start out hearing stuff like that? At NUMMI, the message when we came aboard was "Welcome to the family."

Team Leader

Once you start working as a real team, you're not just work acquaintances anymore. When you really have confidence in your co-workers, you trust them, you're proud of what you can do together, then you become loyal to them. That's what keeps the absenteeism rate so low here. When I wake up in the morning, I know there's no one out there to replace me if I'm feeling sick or hung over or whatever. . . . At NUMMI, I know my team needs me.

Team Leader

The average worker is definitely busier at NUMMI than he was at Fremont. That's the point of the NUMMI production system and the way it ties together standardized work, no inventories, and no quality defects. The work teams at NUMMI aren't like the autonomous teams you read about in other plants. Here we're not autonomous, because we're all tied together really tightly. But it's not like we're just getting squeezed to work harder, because it's the workers who are making the whole thing work— we're the ones that make the standardized work and the *kaizen* suggestions. We run the plant—and if it's not running right, we stop it. At GM-Fremont, we ran only our own little jobs. We'd work really fast to build up a stock cushion so we could take a break for a few minutes to smoke a cigarette or chat with a buddy. That kind of "hurry up and wait" game made work really tiring. There was material and finished parts all over the place, and half of it was defective anyway. Being consistently busy without being hassled and without being overworked takes a lot of the pain out of the job. You work harder at NUMMI, but I swear it, you go home at the end of the day feeling less tired—and feeling a hell of a lot better about yourself!

Team Member

In our standardized work training, our teachers told us we should approach our fellow team members and suggest ways to improve their jobs. Hell, do you see me trying that with a team member who's six-foot-four and weighs 250 pounds? You'd be picking me up off the floor if I tried that. . . . Standardized work is a joke as far as I can see. We're supposed to go to management and tell them when we have extra seconds to spare. Why would I do that when all that will happen is that they'll take my spare seconds away and work me even harder than before? I'd rather just do the job the way I'm already comfortable with. I'm no fool.

Department Manager

Our assumption at NUMMI is that people come to work to do a fair day's work. There are exceptions, and you would be foolish to ignore them. But 90% of people, if you give them a chance to work smarter and improve their jobs, and if they find that by doing that they have created free time for themselves, will spontaneously look for new things to do. I've got hundreds of examples. I don't think that people work harder at NUMMI than in other plants. Not physically anyway. But the mental challenge is much greater.

Team Leader

I don't think industrial engineers are dumb. They're just ignorant. Anyone can watch someone else doing a job and come up with improvement suggestions that sound good. . . . And it's even easier to come up with the ideal procedure if you don't even bother to watch the worker at work, but just do it from your office, on paper. Almost anything can look good that way. Even when we do our own analysis in our teams, some of the silliest ideas can slip through before we actually try them out.

There's a lot of things that enter into a good job design. . . . The person actually doing the job is the only one who can see all factors. And in the United States, engineers have never had to work on the floor—not like in Japan. So they don't know what they don't know. . . . Today we drive the process, and if we need help, the engineer is there the next day to work on it with us.

UAW Official

One thing I really like about the Toyota style is that they'll put in a machine to save you from bending down. The Toyota philosophy is that the worker should use the machine and not vice versa. . . . It would be fine if the robots worked perfectly—and the engineers always seem to imagine that they will. But they don't, so the worker ends up being used by the machine. At NUMMI, we just put in a robot for installing the spare tire—that really helps the worker, because it was always a hell of a tiring job. It took awhile, and we had to raise it in the safety meetings and argue about it. And they came through. That would never happen at GM-Fremont—you never saw automation simply to help the worker.

UAW Official

In the future we're going to need union leaders with more technical and management knowledge. We're much more involved now in deciding how the plant operates. That stretches our capabilities. Management is coming to us asking for our input. . . . The old approach was much simpler—"You make the damned decision, and I'll grieve it if I want." Now we need to understand how the production system works, to take the time to analyze things, to formulate much more detailed proposals. This system really allows us to take as much power as we know what to do with.

UAW Official

Now when I try to explain [NUMMI] to old UAW buddies from other plants . . . they figure that I'm forced to say all this stuff because they shut our plant down and I had no choice. They figure going along with the team concept and all the rest was just the price we had to pay to get our jobs back. I explain to them that the plant is cleaner, it's safer, we've got more say on important issues, and we have a real opportunity to build our strength as a union. I explain to them that our members can broaden their understanding of the manufacturing system and build their self-esteem, and that the training we've gotten in manufacturing, problem solving, quality, and so on can help them reach their full potential and get more out of their lives. I explain to them that in a system like this, workers have got a chance to make a real contribution to society—we don't have to let managers do all the thinking. But these guys just don't see it. Maybe it's because they haven't personally experienced the way NUMMI works.

Whatever the reason, they just see it all as weakening the union. Someone like Irving Bluestone probably understands what we're doing. He had the idea a long time ago: if the worker has the right to vote for the president of the United States, he ought to have the right to participate in decisions on the shop floor.

Team Member

In the old days, we had to worry about management playing its games, and the union was there to defend us. But now, with the union taking on its new role, it's not as simple as before, and we have to worry about both the management games and the union games. I don't want the type of union muscle we used to have. You could get away with almost anything in the old plant, because the union would get you off the hook. It was really crazy. But it wasn't productive.

Team Leader

There are people here who will tell you they hate this place. All I say is: actions speak louder than words. If people were disgruntled, there's no way that we'd be building the highest quality vehicle. You wouldn't have a plant that's this clean. You would still have the drug problems we had before. You would still have all the yelling and screaming. You can't force all that. And try this: go into any of the bathrooms, and you'll see there's no graffiti. If people have a problem with their manager, they don't have to tell him on the bathroom wall. They can tell him to his face. And the boss's first words will be: "Why?" Something's happened here at NUMMI. When I was at GM, I remember a few years ago I got an award from my foreman for coming to work for a full 40 hours in one week. A certificate! At NUMMI, I've had perfect attendance for two years.

Other things being equal, work that is intrinsically motivating—as opposed to mundane and routine—is better than work that isn't. But workers at NUMMI recognize that other things are *not* equal, and they are realistic in their recognition of having had an unlucky draw in terms of education and opportunity. They see automobile assembly as work that can never have much intrinsic value, but they understand that their own motivation levels can nevertheless vary from strongly negative, at GM-Fremont, to strongly positive, at NUMMI.

"What we have here is not some workers' utopia," said one NUMMI worker. "Working on an assembly line in an automobile factory is still a lousy job. . . . We want to continue to minimize the negative parts of the job by utilizing the new system." Even though this work lacks the kind of intrinsic interest that would bring a worker in on a free Sunday, for example, the difference between the levels of motivation at NUMMI and at GM-Fremont spells the difference between world-class and worst-in-class.

The third explanation of increased motivation is the respect and trust that management shows workers in NUMMI's ongoing operations. For example, when the plant first began operations, the new NUMMI managers responded quickly to requests from workers and union representatives for items like new gloves and floor mats, which surprised workers used to seeing requests like these turn into battles over management prerogative.

After a few months of getting everything they asked for, workers and union representatives started trying to think of ways to reciprocate. Eventually, they decided that chrome water fountains were unnecessary and told management they'd found some plastic ones for half the price. A few weeks later, management upped the ante one more time by giving work teams their own accounts so they could order supplies for team members without prior approval from management. This kind of behavior led workers to conclude that they did indeed share common goals with management.

Power and Empowerment

The NUMMI production system confronts us with a set of formalized procedures that seem designed not primarily as instruments of domination but as elements of productive technique that all participants recognize as tools in their own collective interest. Management *and* labor support the NUMMI system. In fact, the first and overwhelming fact to emerge from interviews is that no one at NUMMI wants to go back to the old GM-Fremont days. Whatever their criticisms and whatever their positions, everyone feels that NUMMI is a far superior work environment.

NUMMI's no-layoff policy, management efforts to build an atmosphere of trust and respect, the NUMMI production system—especially the stimulus of its learning orientation—all help to explain this attitude. Beyond these formal policies, however, there are two more factors that help explain NUMMI's success with workers. The first of these, as we've seen, is the psychology of work. The final piece of the puzzle has to do with power.

There are two kinds of power to consider: hierarchical power within the organization and the power balance between labor and management. NUMMI takes a distinctive approach to both.

In terms of hierarchical layers, NUMMI is a fairly typical U.S. manufacturing plant, and in this sense, as well as in work-flow procedures, it is a very bureaucratic organization. NUMMI's structure is not flat. It has several well-populated layers of middle management. But consistent with the idea of turning the technologies of coercion into tools for learning, the function of hierarchy at NUMMI is not control but support.

Decisions at NUMMI are made by broad vertical and horizontal consensus. At first glance, decision making appears to be somewhat *more* centralized than at most U.S. factories, but this is because consensus-based decision making draws higher and lower layers into a dialogue, not because higher levels wield greater unilateral control. Both ends of the hierarchical spectrum are drawn into more decision-making discussions than either would experience in a conventional organization.

The contrast with the popular approaches to empowerment is striking. At one U.S. telecommunications company, the model organization today is a plant of 90 workers in self-managed teams, all reporting to a single plant manager. The company's old model included a heavy layer of middle management whose key function was to command and control, so it is easy to understand the inspiring effect of the new approach. But at NUMMI, middle management layers are layers of expertise, not of rights to command, and if middle managers have authority, it is the authority of experience, mastery, and the capacity to coach.

As for the second aspect of power, many observers have assumed that the intense discipline of Toyota-style operations requires complete management control over workers and elimination of independent work-force and union power. But at NUMMI, the power of workers and the union local is still considerable. In some ways, their power has actually increased. In fact, it may be that the NUMMI model has succeeded only *because* of this high level of worker and union power.

What makes the NUMMI production system so enormously effective is its ability to make production problems immediately visible and to mobilize the power of teamwork. Implemented with trust and respect, both these features of the system create real empowerment. Wielded autocratically, they would have the opposite effect. Visible control could easily turn into ubiquitous surveillance. Teamwork could become a means of mobilizing peer pressure. A healthy level of challenge could degenerate into stress and anxiety.

The NUMMI production system thus gives managers enormous potential control over workers. With this potential power ready at hand, and under pressure to improve business performance, there is a real danger that the relationship will sooner or later slide back into the old coercive pattern.

But such a slide would have an immediate and substantial negative impact on business performance, because labor would respond in kind. An alienated work force wipes out the very foundation of continuous improvement and dries up the flow of worker suggestions that fuel it. And the lack of inventory buffers means that disaffected workers could easily bring the whole just-in-time production system to a grinding halt. Alongside workers' positive power to improve quality and efficiency, the system also gives workers an enormous negative power to disrupt production.

In other words, NUMMI's production system increases the power both of management over workers and of workers over management.

A system this highly charged needs a robust governance process in which the voices of management and labor can be clearly heard and effectively harmonized on high-level policy issues as well as on work-team operating issues. The union gives workers this voice.

When, for example, workers felt frustrated by what they saw as favoritism in management's selection of team leaders, the union largely eliminated the problem by negotiating a joint union-management selection process based on objective tests and performance criteria.

As one UAW official put it, "The key to NUMMI's success is that management gave up some of its power, some of its traditional prerogatives. If managers want to motivate workers to contribute and to learn, they have to give up some of their power. If managers want workers to trust them, we need to be 50-50 in making the decision. Don't just make the decision and say, 'Trust me.'"

Union leaders and top management confer regularly on- and off-site to consider a broad range of policy issues that go far beyond the traditional scope of collective bargaining. The union local has embraced the NUMMI concept and its goals. But its ability and willingness to act as a vehicle for worker concerns adds greatly to the long-term effectiveness of the organization.

NUMMI's ability to sustain its productivity, quality, and improvement record now depends on workers' motivation, which rests, in turn, on the perception and reality of influence, control, and equitable treatment. It is in management's own interest that any abuse of management prerogatives should meet with swift and certain penalties. The contribution of labor's positive power depends on the reality of its negative power.

In this way, the union not only serves workers' special interests, it also serves the larger strategic goals of the business by effectively depriving management of absolute domain and helping to maintain management discipline.

Empowerment is a powerful and increasingly popular approach to reinvigorating moribund organizations. The NUMMI case points up two of empowerment's potential pitfalls and suggests ways of overcoming them.

First, worker empowerment degenerates into exploitation if changes at the first level of management are not continuously reinforced by changes throughout the management hierarchy. Strong employee voice is needed to ensure that shop-floor concerns are heard at all levels of management. Without it, workers' new power is little more than the power to make more money for management.

Second, worker empowerment degenerates into abandonment if work teams fail to get the right tools, training in their use, and support in their implementation. Standardized work, extensive training in problem solving, a responsive management hierarchy, and supportive specialist functions are key success factors for empowerment strategies.

Taylorist time-and-motion discipline and formal bureaucratic structures are essential for efficiency and quality in routine operations. But these principles of organizational design need not lead to rigidity and alienation. NUMMI points the way beyond Taylor-as-villain to the design of a truly learning-oriented bureaucracy.

4

Trial-By-Fire Transformation: An Interview with Globe Metallurgical's Arden C. Sims

Bruce Rayner

Globe Metallurgical Inc. is well known as the first small company to win the Malcolm Baldrige National Quality Award in 1988. It is less well known for its management-led leveraged buyout, flexible work teams, strong R&D focus, and a high-value-added niche marketing strategy. Taken together, these innovations have transformed the $115 million company from an old-fashioned, rusty supplier of commodity metals, selling to the ailing steel industry, into the preeminent source for specialty metals to the chemical and foundry industries worldwide.

Globe's eight-year transformation did not come about by hiring consultants or listening to experts. It came about in the day-to-day crucible of managing. Faced with a series of crises, Globe found what worked—and what worked are the innovations that today make cutting-edge companies.

Leading the company through the changes was chief executive Arden C. Sims, the slow-talking son of a West Virginian coal miner. When Sims was hired in March 1984, Globe was an inconsequential division of Interlake Corporation, a large steel conglomerate headquartered at the time in Oak Brook, Illinois. Globe had more than its share of problems: rising power rates, rigid work rules imposed by the United Steelworkers union, outdated production technology, and fierce competition from low-cost foreign suppliers. To add insult to injury, three months after he was hired, Interlake's management told Sims that it planned to sell Globe.

Sims's initial approach to Globe's problems was traditional reactive management: cut costs to regain competitiveness. But the combination of slashing salaried positions and revamping operations proved inadequate against the new competitive realities. Backed into a corner, Sims began experimenting. At the prompting of customers, he attended a seminar on total quality, paving the way for the company's quality program; he discovered the power of flexible work teams when management was forced to run the furnaces during a year-long strike at one of the company's two plants; he organized an LBO that allowed for even more dramatic changes in the work order; and he took the company global and into highly profitable niche markets after severing a 34-year relationship with Globe's sales and marketing representative.

In all of these instances, the innovations were serendipitous. But once discovered, Sims skillfully turned them to Globe's advantage and, in the process, reinvented the company. Globe now leads its industry in virtually all performance measures—cost, productivity, quality, and sales and profit growth—and has won numerous quality awards besides the Baldrige. Currently, Globe operates between a 90% and 100% capacity utilization rate, compared with 40% for the rest of the industry.

The company has two production facilities, one in Beverly, Ohio and one in Selma, Alabama. The Beverly plant makes specialty ferroalloy products used by the foundry industry in the production of such things as camshafts, crankshafts, and brake housings. The Selma plant produces silicon metals, which are purchased by the chemical industry to manufacture high-grade silicone, used in everything from computer chips to aerosol spray, to plastic bottles. This interview was conducted at Globe's Beverly, Ohio plant by HBR associate editor Bruce Rayner.

HBR: When you applied for the Baldrige Award, did you expect to win?

Arden C. Sims: Frankly, we didn't even expect to apply. None of Globe's managers had heard of the award until Ken Leach, the vice president of administration, who was also in charge of quality, picked up an application in May 1988. He asked me what I thought about us applying. I looked the application over and said it looked like a pretty good idea to me. That was a week before the deadline for entries, so a few of us sat down over the weekend and wrote out the application on an Apple Macintosh—all 63 pages—and sent it off on Monday.

We pretty much forgot about it after that until I got a call over the summer to say we would be getting a site visit in September. Then we grew more interested. The examiners came to our Beverly plant, looked around for a couple of days, and left. A few weeks later, I got

Exhibit I.

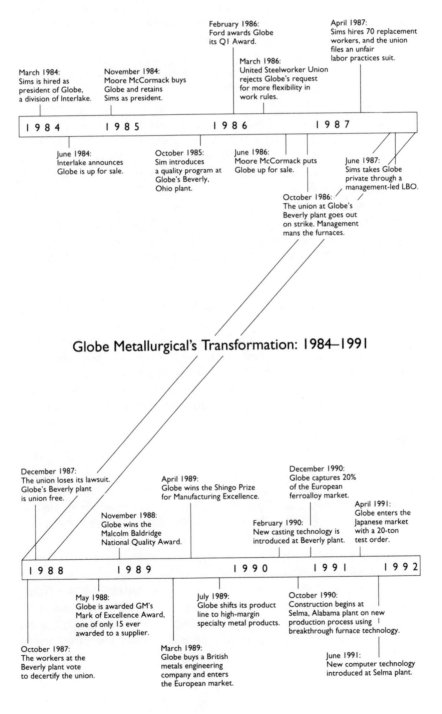

Globe Metallurgical's Transformation: 1984–1991

March 1984:
Sims is hired as president of Globe, a division of Interlake.

November 1984:
Moore McCormack buys Globe and retains Sims as president.

February 1986:
Ford awards Globe its QI Award.

March 1986:
United Steelworker Union rejects Globe's request for more flexibility in work rules.

April 1987:
Sims hires 70 replacement workers, and the union files an unfair labor practices suit.

1 9 8 4   1 9 8 5   1 9 8 6   1 9 8 7

June 1984:
Interlake announces Globe is up for sale.

October 1985:
Sim introduces a quality program at Globe's Beverly, Ohio plant.

June 1986:
Moore McCormack puts Globe up for sale.

June 1987:
Sims takes Globe private through a management-led LBO.

October 1986:
The union at Globe's Beverly plant goes out on strike. Management mans the furnaces.

December 1987:
The union loses its lawsuit. Globe's Beverly plant is union free.

April 1989:
Globe wins the Shingo Prize for Manufacturing Excellence.

December 1990:
Globe captures 20% of the European ferroalloy market.

April 1991:
Globe enters the Japanese market with a 20-ton test order.

November 1988:
Globe wins the Malcolm Baldridge National Quality Award.

February 1990:
New casting technology is introduced at Beverly plant.

1 9 8 8   1 9 8 9   1 9 9 0   1 9 9 1   1 9 9 2

May 1988:
Globe is awarded GM's Mark of Excellence Award, one of only 15 ever awarded to a supplier.

July 1989:
Globe shifts its product line to high-margin specialty metal products.

October 1990:
Construction begins at Selma, Alabama plant on new production process using breakthrough furnace technology.

October 1987:
The workers at the Beverly plant vote to decertify the union.

March 1989:
Globe buys a British metals engineering company and enters the European market.

June 1991:
New computer technology introduced at Selma plant.

a call from Secretary of Commerce William Verity. He said, "I've got some good news and some bad news." The good news was we'd won, and the bad news was I couldn't tell anyone until the official announcement. It was a long few weeks.

So unlike a lot of companies, Globe did not use the Baldrige Award application as a way to learn about total quality?

No. The Baldrige Award was more of a reward for our efforts than a learning experience. We had been pursuing quality pretty steadily since late 1985. Our backs were up against the wall, we were on the defensive, and I saw total quality as a way to save the company from going out of business.

Of course, quality didn't solve all our problems. It just got us moving in the right direction. It opened my eyes to a host of innovative ideas and practices that changed the way I thought about running the business. It set the stage and helped me break out of our defensive posture and go on the offensive.

What were some of the other elements that played a role in transforming Globe?

I've initiated three other major changes over the last eight years that have fundamentally transformed how the company operates. First, we had a strike in 1986 and 1987 that led to a new work structure that has dramatically increased our productivity. Second, the LBO in mid-1987 allowed me to take greater managerial control of the company's day-to-day operations and started us on the road to greater employee empowerment and involvement. And third, in 1989, we repositioned Globe's product line into more lucrative high-margin specialty products, and we entered foreign markets. To do that, I first had to terminate a 34-year relationship with Pickands Mather, the company that handled all Globe's sales and marketing. All of these pieces had to work together. They all built on our earlier efforts in quality and cost control.

Before joining Globe, I had no experience with quality or work teams or empowerment or LBOs. But with Globe on the ropes, I realized that the old-fashioned approach to cutting costs—adding more supervisors and controls or laying off workers—just wouldn't work anymore. At the time, I didn't know what would replace the old ways. In some cases, I found the solution myself, in other cases, someone directed me there, and in still other cases, the solution evolved over time through trying one thing, then trying something else.

Every time I came across a new idea that I thought would help improve the company's operations, I would adapt it and drive it into the organization until it became part of the company. Over time, each change contributed to improving our competitive position until we found that we had created a brand new company. We've gone from worst practice to best practice and from low-tech to high-tech. It's as if the company was near death, and I found the medicine that would restore it to full health. I didn't invent the medicine, I just administered it and stood by as the patient recovered.

Let's take each of the changes you made in turn, starting with quality. Why did quality become so important to Globe in 1985?

To understand how Globe came to quality, you have to first understand where the company was when I joined it in 1984. Globe was an archetypal rust-belt company: old-fashioned, bureaucratic, and unresponsive, with high costs and low quality. We were in a death spiral. If things had simply gone on with business as usual, we'd have been put out of business by foreign competitors in a few years. So when we adopted total quality at the end of 1985, it was a reaction to multiple crises, both internal and external, that had been brewing for years.

I was hired in March 1984 by Globe's parent, Interlake Corporation, because Globe's president was retiring and Interlake wanted a younger man who could turn things around. The former president had been with the company in one capacity or another since 1952, and he made all the day-to-day decisions. Even though he had a very talented management team, all the engineering, metallurgical, and operating decisions funneled up to him. I know for a fact that you can't run a company very effectively with one person making all the decisions. To make matters worse, a lot of the budget-related decisions had to go through Interlake's management. That meant further delays and more bureaucracy; approval for any kind of capital spending took months. But Interlake wasn't investing much. As far as I could tell, Interlake management had been using Globe as a cash cow with little concern for the company's future. We had no research and development, the product line hadn't changed in years, and plant and equipment were aging.

What was your mandate when you were hired?

As I said, I was hired to turn things around. But it wasn't long before I found out that Interlake's management had other ideas. Three months after I joined the company, Interlake put Globe up for sale. This was a complete surprise because I had explored the possibility of Interlake selling Globe before I was hired, and they had assured me the company was secure. Fortunately, I kept my job when Interlake sold Globe in November to Moore McCormack Resources, another steel-related conglomerate.

Moore McCormack bought Globe because we fit strategically with Pickands Mather, which was owned by Moore McCormack at the time. The problem was, Pickands Mather had total control over our sales. It determined what types of metals we sold and who we sold to. It was a difficult relationship, to say the least.

I was doing battle on other fronts as well. Our costs were high, especially power and labor costs. Power rates had recently gone up, and they were going up again. Electricity is a major cost in this business. Globe had no leverage with the utilities, they were walking all over us. Both labor and capital productivity were low because of the United Steelworkers' complicated work rules. We had just about the lowest productivity in the industry, compared with our domestic competitors.

Then there was the offshore competition from companies in Brazil and Argentina. Their prices were so low I suspected them of dumping. I even went to Washington to lobby for some controls on imports in 1985, but I didn't get anywhere. Finally, many of our customers in steel and aluminum were going under or cutting production. They were canceling orders right and left. In the early 1980s, over half of the 6,000 foundries in this country went out of business.

What were some of the first steps you took when you joined Globe?

I started making changes as soon as I was hired. I cut the salaried work force by 13, restructured the engineering department and the accounting system, and computerized the furnaces and some of the other processes in the plant. But all this wasn't enough to pull Globe out of its downward spiral. It became clear to me that the old approach to controlling costs had serious limitations; no matter how much I cut and restructured, Globe would never be competitive unless we learned to control the production process more exactly. I was looking for anything that might help.

About this time, some of Globe's major customers—Ford, GM, Intermet—suggested I learn about total quality, so I arranged to attend a week-long quality seminar in October 1985. It was my first exposure to the concepts of quality as something more than just postproduction inspection. I figured if we went about it the right way, a full-blown quality program at Globe could reduce costs and improve performance by a few million dollars.

A lot of companies approach quality as a long-term cultural change process, but quality at Globe was a way to cut costs and cut them as quickly as possible.

That's exactly right. Our quality program was not about trying to change the company's culture, it was about digging our way out of a real crisis. There was nothing touchy-feely about our commitment to quality.

Even the name of our quality program indicates where our priorities lay: we called it QEC, for quality, efficiency, and cost. QEC was, and still is, about gaining control of our processes, improving the efficiency of the furnaces and the consistency of the product in order to get out excess cost. At Globe, controlling the process is especially critical because we run our furnaces 24 hours a day, 7 days a week. Any variation in the parameters that affect the process produces a variation in the product, and that's not acceptable to our customers.

Can you calculate the annual cost savings from improved quality?

Sure. We had to make the calculation when we won the Shingo Prize in 1989. Between 1986 and 1988, Globe eliminated $11.3 million of operating cost, and most of it was a result of our quality programs. That was a 367% cost reduction over three years. We have maintained the same pace for the last few years, cutting about $4 million a year out of costs through continuous quality improvement. Right now, we are the most efficient producer in the whole industry.

By controlling the process more exactly, we have also improved the operating efficiency and power consumption of our furnaces by about 30% since 1984—that saves us an additional $3 million to $5 million a year. We still have another 20% to 30% to go before we will be operating at our theoretical maximum efficiency, and that doesn't include the improvements we will make when our new process technology comes on-line. Including all the innovations we are planning, I expect our annual cost savings to increase to about $13 million a year by 1995. That's significantly more than my initial expectation of cutting a few million dollars. If you had told me back in 1985 that we would be able to get these kinds of savings from quality, I never would have believed you. It shows you how powerful the techniques can be.

You introduced your quality program at the Beverly, Ohio plant first. Was that a pilot program? Did you want to get the bugs out before implementing it at the Selma, Alabama plant?

Not at all, there was nothing "pilot" about it. We introduced a full-fledged quality program at the Beverly plant because that was where we had the most serious operational problems. It's older than our Selma plant and was where all the salaried work force was located—accounting, engineering, personnel. My office was at Beverly too, so it made it easier to oversee the program's rollout.

The first thing we did was give all the plant's employees, union and salaried workers, six hours of training in basic statistical process control, control charts, how to use a calculator, those sorts of things. It was a pretty serious session. We told them that quality was not an option, it was a requirement. They knew the company was in trouble and that we needed to make some changes. We tried to make it clear that they either accept the program or there wasn't a place for them at the company.

My impression was that they listened carefully, but I'm not sure that they were totally convinced that this was going to be with us long term. They were used to a new program every 6 or 12 months, so naturally there was skepticism. They all attended the classes, they paid attention, but I'd have to say that it wasn't accepted by everyone.

As soon as we applied the techniques in the plant, though, they realized this was serious. If a manager went out on the shop floor for an inspection and found that a supervisor hadn't filled in his control charts for that shift, we'd call him into the office. I'd tell him if he didn't fill in the chart every time, as we'd asked, he'd lose his job. They knew I meant it because I'd already laid off some of the salaried work force. Fortunately, I didn't have to fire anyone. They all shaped up pretty quick.

You say that your customers were also driving you toward quality. How influential were they in shaping Globe's quality program?

Very influential because the pressure from our customers was real. All of them were cutting the number of suppliers they dealt with, so we had to improve our quality or we wouldn't be supplying them anymore. Needless to say, we were very open to all their suggestions.

One of the first companies we worked with on quality was Ford. A Ford team audited us for the Q1 Award in late 1985, soon after we introduced our QEC program. We missed the Q1 by only a few points. They suggested a few adjustments to our program, such as changes in our measurement system, and we made them. When the audit team came back in early 1986, we qualified comfortably. It's been that way with our other quality-conscious customers too: whatever they want us to do—documentation, quality techniques—we do it.

But there is more to getting the customer's order than having a quality certificate. You have to be able to deliver a consistent product every time. To do that, we needed more than just good process control. We needed to make sure that our frontline employees understood the importance of quality.

So in 1986, I started Globe's customer-visit program for our Beverly plant employees to drive home the importance of quality and consistency. The idea came to me in late 1985 when I visited a GM foundry and saw workers adding a cupful of our product to a mold of molten metal. That really impressed me. I thought, Globe ships tons of that product a year, but here was GM adding a cupful at a time. Our product had to be of the highest quality and very consistent or else their production run would be ruined. I thought, if the plant workers could see that firsthand, they'd realize how important their job was to the customer.

Did it work?

Yes and no. The first visit was to the Ford plant in Cleveland in 1986. We sent about 30 workers, both union and supervisors. They all came back with a different attitude about their jobs and a greater appreciation for the customer's requirements.

But we couldn't seem to translate that new awareness into improved performance. The problem was the union's rigid work rules. Without greater flexibility and worker involvement on the plant floor, we couldn't improve productivity and move ahead with our quality program. The union members were willing to fill in control charts, but they were not committed to continuous improvement. Understandably, they felt that continuous improvement was in direct conflict with the preservation of their jobs.

What was the solution?

Renegotiate the contract. Our productivity was lowest at the Beverly, Ohio plant with the United Steelworkers union. The union's contract was coming up for renewal in October 1986, so I saw that as a chance to make some changes. Normally the negotiations would begin in late September, but I requested a meeting with the local's senior management in March. Also, the union work force was scheduled for a 6% pay hike in April. I wanted to postpone the pay raise and negotiate changes in the work rules as soon as possible, before the normal negotiations started.

Surely the plant's union wasn't about to give up a 6% pay raise?

No, they didn't make any concessions at all. In fact, the union president didn't even show up for the meeting—only the vice president, treasurer, and secretary. Still, we opened Globe's books to them and explained how serious the situation was—we were in the red and had been for quite a while. We explained the cost advantage that countries like Brazil had over us in electricity rates, labor rates, and raw materials. We even compared our labor costs with our three main domestic competitors, most of which had lower wage rates and higher productivity rates. We argued that if we couldn't find a solution, we would all be out of work soon. The bottom line was that the cost had to come out of either the number of employees or out of wages and benefits to the tune of about $1.5 million. That was equivalent to about 40 full-time workers.

The union stonewalled us. They said a contract is a contract, and they'd negotiate a new one in September. That was the end of the discussion.

While I was trying to decide on my next move, circumstances changed dramatically. In June 1986, LTV, Moore McCormack's largest steel customer representing about 25% of its total revenue, went into Chapter 11. It was a real blow to Moore McCormack. Globe was once again put up for sale along with all of Moore McCormack's other metals-related businesses, including its ore and coal mines, and its shipping lines.

Soon after Moore McCormack announced Globe was up for sale, we had another meeting with the union to discuss our predicament. We got the same reaction: no deals. One union representative even said he didn't think the company was really for sale. He said it was just a ploy to try to negotiate concessions from the union. I was amazed by that statement. It showed just how far apart we really were. That was the last meeting until the contract negotiations began for real in late September. They must have figured they were in a pretty good bargaining position; no way would management risk a strike with the company up for sale.

But they were wrong?

Yes. In late August, we started preparations for a strike. I brought in a security consultant and put up a fence around the property. As the contract's expiration date got closer, I made arrangements to have food and bedding brought in for the replacement workers, management and salaried employees who would be working in the plant if a strike did occur. It was a precaution, just in case we couldn't get in and out of the plant. We arranged for most shipments to come by rail instead of by truck to reduce the chance of violence at the picket line, and we secured a warehouse about 20 miles away on the rail line for packaging and shipping finished goods.

But even though we prepared for a strike, I fully expected to reach a settlement. I couldn't imagine the plant without a union work force. In September, the union made a move toward some of our demands. They set up a team that met with a management team and jointly agreed to eliminate 15 or 20 jobs. But that wasn't enough, I wanted 40.

As the deadline approached, we'd go back and forth. One day it looked hopeless, the next it looked encouraging. The day before the deadline, there was some movement and I felt that if the union membership had a chance to look at the deal we'd come up with, it might pass. But the union negotiating committee, headed by the USW's regional representative, decided not to put it up for a vote by the plant's members. I still think that was a big mistake on the union's part. They did their members a real disservice.

What was your reaction when there was no vote?

I knew there was going to be a strike. We agreed to a week's extension, but nothing happened, they were just buying time. So at the end of the shift at 11:00 P.M. on October 8, 1986, the union walked out of the plant and the pickets went up at the front gate. As the union workers left the plant, about 35 salaried workers and 10 company managers stepped in to take over operation of two of the five furnaces. We shut the other three furnaces down.

I was assigned to work on the maintenance crew, the dirtiest job in the whole plant. I still don't know who made the assignments. That first night, I helped set up the kitchen. It was our job to install the stoves, refrigerators, freezers, sinks, and other equipment. We also put up a satellite dish. The next morning, as soon as it was daylight, we went out in the plant to start the maintenance work. The plant was quite a mess, metal had been spilled on the floor, and the equipment was in poor repair. We brought in picks and shovels and front-end loaders and started cleaning up around the furnaces.

We stayed in the plant for three days and two nights. On the third day, we sent a few people out to test the picket line to see if we could get in and out OK. There was no trouble, so we took turns going home and getting cleaned up. After that, we settled into a regular routine of 12-hour shifts, 7 days a week, and continued that way pretty much for the next six months. Sometimes we'd work longer, depending on what we needed to do each day, but everyone was scheduled 12 on, 12 off.

Did the strike ever turn ugly?

We had a few incidents. It turned nasty when we started bringing some raw materials in by truck instead of rail. A truck was attacked, the strikers broke the windshield and damaged the truck. The driver was OK. The police came up, and we got an injunction that cut the number of pickets down to four. After that, the union set them up about two miles from the plant in a sharp bend in the road. We had a few incidents there too, a worker's truck was overturned.

I had security people guarding my house for about three years as a result of the strike. The first year it was round-the-clock protection. I never had any incidents there, but we had lots of people come by to take a look. Overall, there was a lot of tension in the community. Beverly is a pretty small town, and the plant was the community's biggest employer. There'd be confrontations at bars, sporting events, any public gathering. It was pretty uncomfortable.

What did the experience of working in the plant teach you?

The strike was a time of great stress but also a time of great progress. We experimented with everything. And as we did, we began to see all kinds of new possibilities in the way work was performed because we didn't have to worry about honoring job classifications or work rules. Our objective was to find the most efficient way to run the furnaces, with no constraints on how we did it. A few weeks after management took over operating the plant, output actually improved by 20%.

We were operating in a very fast, continuous improvement mode. Every day, people would suggest ways to improve the operation of the furnaces or the additive process or the way we transported material around the plant. I kept a pocket notebook, and if I saw something, I'd jot it down and discuss it with the team over coffee or during a meal. I filled a notebook every day.

A lot of new practices came from our informal discussions in the kitchen. I remember one mealtime, we were talking about the process of breaking up the metal after it had set in the cooling molds—large cast-iron trays about eight feet across and eight inches deep. The metal was transferred from the mold and passed across a grid where someone would break it up manually. As the metal fell through the grid, it landed in containers that were then transported by forklift to the bulk storage area. The managers working on the grid came in one day and said they thought it was an unnecessary step. They proposed dumping it directly from the mold into a truck, weighing it, and then dumping it in the storage pile. Just the action of dumping the metal from the mold to the truck was enough to break it up sufficiently. We tried it, it worked, and so we eliminated two jobs around the clock, or eight workers. That was a savings of more than $300,000 a year. People made those kinds of suggestions constantly. By July 1987, when the strike was in its tenth month, Globe was making a profit.

As we made more and more changes and as we settled in to the routine of running the plant, it became evident that we didn't need first-line supervisors. We could produce the product more effectively if everyone just worked together cooperatively—welders, crane operators, furnace operators, forklift drivers, stokers, furnace tappers, and tapper assistants.

You mean you developed cross-functional teams?

Yes, but that's not what we called them. We had a shift of eight men running a furnace, each with a group leader, and they could all do each other's jobs. We didn't start out with the idea of eight-man teams, it just happened that was the best way we found to do the work. With that kind of flexibility, we were able to get the number of shop-floor workers down from about 350 people before the strike to about 120 after. That's how we operate today.

You had initially asked the union to cut 40 full-time-equivalent jobs, but you discovered that you could cut more than five times that number while, at the same time, improving productivity. When you realized you could make such dramatic changes, did you change your negotiating tactics?

My attitude certainly changed, but through early 1987, I thought an agreement was desirable because we were up for sale. As time passed, though, I grew less willing to negotiate. In April, I made a decision to convert the 35 plant supervisors, who were salaried employees, to full-time hourly workers. I also hired about the same number of new workers, all nonunion, many of them farmers and none with any experience around furnaces. Once I had hired replacements, I knew there was no going back to the old ways.

The union responded by filing an unfair labor practices suit with the National Labor Relations Board in Washington, D.C. If they won the suit, they'd be reinstated and entitled to millions of dollars in back pay. That was a big concern. In October 1987, a year after the strike started, the union made an unconditional offer to return to work. I think that was when they realized that they had lost the fight. At the same time, the new workers voted to decertify the union. Then, on Christmas eve, I read in the local paper that the judge had ruled against the union in the lawsuit. So by the end of the year, the strike was over and we had won.

Winning was a great Christmas present, but even better was the stunning improvement in productivity. Early in 1988, I called back about 30 ex-union workers, and we fired up the other three furnaces.

When the strike began in October 1986, did you ever think that you would end up with a nonunion shop?

No. Absolutely not. We didn't start out trying to break the union. As I said, I didn't even realize it was possible to operate the plant without a union because it had always had one. That's just the way it evolved. I don't think a company should ever start out by trying to break a union. But if it reaches a point where you are fighting for survival, as we were, then you have to do whatever's necessary.

In some ways, the most amazing part of the strike is the fact that right in the middle of it all, you led a management buyout of the company. How did that come about? Was it part of your strategy for transforming the company?

Not at first. I never even considered an LBO when Moore McCormack put Globe on the block in mid-1986. I fully expected the company to be sold to another large company. In fact, in the first round of bidding in February 1987, it almost was, but the buyer's board backed down and the company withdrew.

But the more I thought about it and the more I realized what an LBO could mean for the company, the more appealing it became. I saw it as a way to drive change further down into the company, to bolster our total quality efforts by involving the workers more in day-to-day operations. The LBO became a very important piece of the puzzle for transforming the company.

How did you learn about LBOs?

I spent January and February of 1987 preparing a five-year plan for prospective buyers to look at—cost and revenue projections, capital requirements, annual budgets, that sort of thing. I would give the presentation to prospective buyers. In all, about 15 came through, a number of them Wall Street LBO types. After I made the presentation, many of the Wall Street financiers would approach me with their ideas for a leveraged buyout. But all their plans sounded the same: buy the company, break it up, and sell the pieces off for a profit. That didn't interest me. I wanted to build the company.

Only one guy, Jonathan Lee, chairman of Lee Capital, seemed interested in my plans. There was good chemistry between us, he explained the LBO process carefully and didn't try to pull any punches. The more we talked about the possibilities for Globe as a management-owned company, the more excited we both got. And the more Lee got to know Globe, the more serious he got about investing for the long term. We talked about adding volume and reinvesting the cash flow back into research and development, neither of which are common in an LBO situation. I trusted him, so we teamed up and submitted a bid to Moore McCormack.

When the first buyer backed down, Moore McCormack asked us to make another bid. We did, and we took ownership on June 20, 1987.

How did it feel to be the CEO of a small company, without the backing of a cash-rich parent?

It was sobering, to say the least, mainly because of the strike situation. I immediately called a meeting with everyone at the Beverly plant to explain the new situation. I tried to be as open as I could about the advantages and the disadvantages of taking the company private. There was some concern on the part of the employees and some of our customers that if things didn't go well, we might go out of business. A lot of people shared that concern, myself included. I tried to impress on everyone that we were all in this together. We all had a responsibility to build the company.

The first thing I said was, "I want to see every dime that we spend." If we were cost conscious before, we were doubly so now. We began to track cost with a microscope—the performance of the furnaces, raw materials, inventories, receivables, everything.

You said that the LBO was an important piece of the puzzle for transforming Globe, particularly for involving your employees more in day-to-day operations. How did you accomplish this?

In a number of ways. I'm a strong believer in treating employees as responsible adults and in rewarding good work. So immediately after the LBO, I initiated four new policies: a profit sharing plan, full disclosure of all company information to all employees, bottom-up budgeting, and a personal commitment to full employment at the Beverly plant.

The profit sharing program motivates our employees by keeping them financially involved in the company. But unlike most plans, I decided Globe's plan should pay out on a quarterly instead of an annual basis: I wanted the employees to make the connection between their actions and the immediate financial reward.

The information sharing initiative was important in terms of gaining the trust of the employees. I hold quarterly meetings that the employees call state-of-the-company meetings because I meet with everyone in both plants, usually in small groups of 10 to 20 employees. We discuss the quarterly sales and cost figures and how the numbers translate into that quarter's profit sharing payout. I don't keep anything back. Besides the numbers, we also talk about whatever is going on with the company at the time—exports, quality, new hires, changes to the pension plan. It is also a time to hear employees' concerns and to discuss their ideas for change. These meetings are vital to me. It is important that I get their feedback and suggestions firsthand, with no filter, so I know what's really going on.

Right after the LBO, I also gave my word to preserve full employment at the Beverly plant. I felt that making the commitment was important if Globe was going to continue cutting costs. I made it clear that if an employee suggested a change that meant the loss of a job, no one would get laid off. Instead, I would shift the worker to another job at the plant. We have eliminated a number of jobs since I introduced that policy, and I have kept my promise.

You also said you introduced a bottom-up budgeting process. What is that?

The information for our annual budgets comes in from two sources: revenue projections from the sales force and cost and production estimates from the plant floor. On the cost side, the plant manager meets with all plant employees to discuss such things as furnace output and metallurgical capabilities to determine what the plant can produce and at what cost levels. He also collects data on the projected savings from new quality and cost-reduction programs and from employee suggestions.

Both the sales and the cost information flow into the accounting department, and they put it together, and we see if the numbers work out for the year. If they don't quite make it, we send the plant manager back to the shop-floor employees to ask them to find additional savings. We also ask the sales force to adjust the product mix to boost margins. After we have arrived at a satisfactory budget, we revisit the cost-reduction estimates on a monthly basis, so we are constantly updating the budget estimates. This whole process keeps our workers involved and our forecasts accurate.

All our empowerment policies—profit sharing, information sharing, full-employment policy, and bottom-up budgeting—feed off each other to keep the employees motivated. I am convinced that without them, we wouldn't be averaging $3 million to $4 million a year in quality-related cost savings.

But even as you gained control of the company after the LBO, your sales and marketing function was still handled by an outside company, Pickands Mather. How did the LBO change that relationship?

The LBO didn't change the relationship. Pickands Mather had been representing Globe in the market since 1955 and was well established domestically. After the LBO, I figured it was in our best interest to maintain a good working relationship with them. But it became clear by early 1989 that we couldn't resolve our differences. Pickands Mather dealt mainly in the commodity end of the business, and I wanted to take Globe into high-margin specialty products. They wanted to remain domestic, and I wanted to expand overseas.

Now, during this period, Globe was well on its way to becoming a totally new company. We had been through the worst of the crisis with the strike and the LBO, but Pickands Mather was holding us back from finishing the job. To me, Pickands Mather represented the old way of doing business and the old way of thinking. It was preventing me from completing Globe's transformation, and I wasn't about to let that happen.

Of course, we'd have discussions, sometimes heated, about specialty versus commodity and about selling overseas. But because Globe was a smaller company, I always lost. Those discussions ended soon after Pickands Mather's sales and marketing director retired in mid-1988. To make a long story short, we terminated our relationship with them in July 1989 and built up an in-house sales force. That let us get much closer to our customers, and we began tailoring our products and processes to our customers' needs. Soon after that, I dropped virtually all our commodity lines and focused exclusively on specialty products and expanding overseas. Today we sell 97% specialty metals, and our margins are much more comfortable.

Why was globalization so important to Globe's transformation?

It was important because it offered growth potential, especially for our foundry products. Globe's market share in the United States was pretty stable, so our prospects domestically were limited. Remember, we had just become a small company, and we needed to keep capacity running at or near 100% to keep unit cost down and operating margins up. We were battling against competitors that were much bigger, and we were still struggling to survive. Entering foreign markets became a necessary part of that struggle.

From a strategic point of view, globalization fit well with Globe's commitment to quality and customer service. Domestically, we were gaining a reputation as a cost and quality leader, and I felt we were now ready to compete with the best in the world. In fact, I considered the last five years of Globe's organizational transformation as preparation for going overseas. Today we are selling in Singapore, Malaysia, Hong Kong, Thailand, India, New Zealand, Australia, Korea, Taiwan, and Japan.

Of course, things like winning the Baldrige Award help you get name recognition overseas, but it doesn't assure success. Getting your foot in the door takes a lot of work and a lot of time. Then convincing the customer to buy from a $115 million supplier rather than a $2 billion supplier is even tougher—until they test our product.

What has been your experience in breaking into new markets: into Europe and Asia?

We've had an easier time in Europe than in Asia because of our 1989 acquisition of the British company Materials & Methods, the premier metals engineering company in the world. M&M had a good position in the U.K. and in Eastern Europe, but it was weak in Western Europe because of stiff competition from other alloy makers. Globe gave M&M a way to enter Western Europe because the quality and cost of our alloys were very competitive. From a market share of about 1% in 1989, Globe now has 20% of the Western European market.

Getting into Asia has been harder, especially Japan. But cracking Japan is crucial to us because it's the biggest market in Asia—about half the size of the U.S. market. I believe the Japanese will eventually exit specialty metals because Japanese power rates are so high and the processes they use are expensive and inefficient. Because of Globe's high quality and low cost, I think we can be the future first-choice supplier in Japan. We are laying the groundwork for that now, but it's not easy. We've made some strategic mistakes.

Back in 1988, Bob Jenkins, Globe's vice president of sales and marketing, made several trips to Japan to visit the big four ferroalloy makers: Osaka Tokushu Gokin, Toyo Denka, Shin-Etsu Chemical, and Kusa Ka. They supply all the big Japanese foundries like Toyota, Nissan, Kubota, and Komatsu. Bob told them we could supply them with product of equal quality at a much lower price—most of them only produce three or four products, whereas we can produce about 180, virtually a different product to fit every customer's specific requirements. But he didn't get anywhere. They'd listen politely, but then nothing happened. That went on for months before we changed our strategy.

In 1989, we targeted a group of foundries and used local sales agents to get in the door. Again, the introductory discussions were very polite, very friendly, but no followup. Eventually, we did get a bite. One company decided to test our product—I think just to get rid of us. They really didn't believe our product would measure up, but to their surprise, it outperformed the Japanese product. So then we got down to a real discussion, and we are now working on supplying that foundry with a test run of a few tons. The problem is there are hundreds of foundries like that one scattered all around Japan, so gaining market share there is proving incredibly time-consuming. Even so, I expect to ship about 3,000 tons to 20 foundries this year. That's about 7% of the total Japanese market.

Has it been easier getting into other Asian markets?

It's always tough getting in the door, but they are easier than the Japanese. Our Korean business has developed very nicely. For instance, at Pusan Cast Iron, we made our presentation, discussed their alloy needs, and developed a product to fit their process. They ran some tests, and now we have 100% of their business. The whole process took all of two months. Pusan is one of the major Korean foundries; it makes crankshafts, camshafts, and such for Daewoo. It is also pretty typical of the other Asian companies we have dealt with. Once you get talking, they are willing to play and usually are very impressed with our product. In many cases, we displace Japanese suppliers.

You paint a fairly rosy picture for Globe's future. You have the right products, a competitive cost structure, high quality, and a growing international business. What about the obstacles that stand in your way? What keeps you up at night?

The main problem Globe faces today is making sure we can fuel growth. That means staying very focused on day-to-day operations so we have the cash flow to finance our investments. I have to keep my eye on the basics all the time.

Right now, the biggest challenge is to turn some of our R&D projects into commercial ventures. Research and development is a major focus at Globe, and we've spent heavily on R&D since the LBO. Now we need to raise about $22 million for new capital equipment over the next few years, that is, mainly for a new, more efficient furnace. If we are successful, we will be so far ahead of the competition, they won't be able to catch us technically. I'm confident we can do it, but it is a concern. If we are successful, we will be in a good position to handle our growing worldwide demand.

No question about it, our strategy for the 1990s is based on exploiting our technological advantage. We are no longer a static company in a mature industry but a high-technology company trying to adapt to new markets. I think our competitors believe they are in a mature industry and the technology is as good as it's going to get. That kind of attitude will prove fatal, not only for them but also for complacent companies in a lot of other industries as well.

Does Globe represent a model for others to emulate?

In some ways, yes, I think so. I consider myself a fairly conservative manager. But over the last eight years, I've had to take some pretty big risks just to survive, and I've had to fight some pretty fierce battles. There are a lot of companies out there, big and small, that face the same kinds of challenges that we faced in the 1980s—low productivity, high costs, the wrong product line. They have to deal with their problems soon, or they won't be around much longer.

I've found that one of the most important components of a successful transformation is to gain the respect of your work force. I don't mean the workers have to like you, although that helps, but they have to understand that you will stand your ground when it comes to forcing through big changes. To initiate change, you have to be tenacious, and you need to show that you're not afraid of getting a little dirt under your fingernails.

It's a shame, but I think a lot of senior managers have lost their resolve and their ability to face up to hard work. Change is never easy, and there are no special formulas, no quick fixes. You just have to roll up your sleeves and keep working at it without backing down.

About the Contributors

Paul S. Adler is associate professor at the University of Southern California School of Business Administration. He recently edited two collections of essays: *Technology and the Future of Work* and *Usability: Turning Technologies into Tools*. His two-year study of the New United Motor Manufacturing Inc. plant in Fremont, California provides the basis for his article.

Chris Argyris is the James B. Conant Professor Emeritus at the Harvard graduate schools of business and education. He is the author of *Overcoming Organizational Defenses* and several *Harvard Business Review* articles including "Double Loop Learning in Organizations" and "Skilled Incompetence."

Michael Beer is professor of organizational behavior and human resource management at the Harvard Business School. With Russell Eisenstat and Bert Spector, he co-authored *The Critical Path to Corporate Renewal* (Harvard Business School Press). He is currently inventing and researching an institutionalized process for continual organizational renewal.

John Seely Brown is a corporate vice president and chief scientist at Xerox, and director of the Xerox Palo Alto Research Center (PARC).

Ram Charan is a Dallas-based management consultant who advises companies in North America, Europe, and Asia on implementing global strategies. In previous issues of the *Harvard Business Review*, he interviewed General Electric chairman, John Welch, and Citicorp chairman, John Reed.

Peter F. Drucker is the Clarke Professor of Social Science and Management at the Claremont Graduate School in Claremont, California. His research in the field of modern organizations and their management is highly regarded throughout the world. He is the author of more than 25 books, which have been translated into 20 languages, and is a frequent contributor to the *Harvard Business Review*. His article is adapted from his new book, *Post-Capitalist Society*.

Russell A. Eisenstat is an independent consultant specializing in the management of large-scale corporate change and innovation, strategic human resource planning, and improving staff group effectiveness. His most recent book, *The Critical Path to Corporate Renewal*, written with Michael Beer and Bert Spector, received the Johnson, Smith & Knisely Award for New Perspectives on Executive Leadership. Prior to his current position he served on the faculty of the Harvard Business School.

Philip B. Evans is vice president of The Boston Consulting Group in the firm's Boston office. His work focuses on strategy and operating issues affecting the financial services and media industries. He has consulted to major newspapers, magazine and book publishers, and banks throughout the world and directed BCG's internal training programs for several years.

Thomas Gilmore is vice president of the Center for Applied Research in Philadelphia. He is the author of *Making Leadership Change: How Organizations and Leaders Can Handle Leadership Transitions Successfully*.

Robert D. Haas is chairman of the board and chief executive officer of Levi Strauss & Co. and director of Levi Strauss Associates, Inc. Prior to joining Levi Strauss, he was an associate with the management consulting firm of McKinsey & Company. Mr. Haas has served on the boards of a wide variety of community organizations and committees and is currently a trustee of the Ford Foundation and director of the American Apparel Association.

Larry Hirschhorn is principal of the Center for Applied Research in Philadelphia. His most recent book is *Managing in the New Team Environment: Skills, Tools, and Methods*.

Robert Howard is a writer and consultant based in the Boston area, and author of *Brave New Workplace*. A former senior editor of the *Harvard Business Review*, he was responsible for editing articles on organizational innovation and human resource management. His own writing on work, technology, and organizational change has appeared in a variety of publications including *HBR*, the *New York Times Book Review*, and *MIT's Technology Review*.

Ikujiro Nonaka is professor of management at the Institute for Business Research of Hitotsubashi University in Tokyo, Japan. His other *Harvard Business Review* articles include "The New New Product Development" (co-author with Hirotaka Takeuchi) and "Market Research the Japanese Way" (co-author with Johny K. Johansson).

Bruce Rayner is a writer and editor-in-chief of *Military & Aerospace Electronics*, a monthly magazine covering business and technology developments in the aerospace industry worldwide. Previously, as associate editor at the *Harvard Business Review*, he specialized in topics of production and operations management, including the total quality management (TQM) movement. Prior to joining *HBR* Mr. Rayner was managing editor of *Electronic Business*.

Lawrence E. Shulman is a senior vice president of The Boston Consulting Group and director of the firm's Chicago office. He has consulted to a wide variety of industries worldwide, specializing in issues of postmerger integration management and improving responsiveness to customers.

Bert Spector is associate professor of organizational behavior and resource management at Northeastern University's College of Business Administration. He was a co-author of *The Critical Path to Corporate Renewal* (Harvard Business School Press) with Michael Beer and Russell Eisenstat.

George Stalk, Jr., recently identified by *Business Week* as one among a new generation of leading management gurus, is senior vice president of The Boston Consulting Group in the firm's Toronto office. His previous publications include *Competing Against Time*, with Thomas Hout, and, with James C. Abegglen, *Kaisha: The Japanese Corporation*. He frequently contributes to the *Harvard Business Review* and other business publications.

William Taylor, former associate editor of the *Harvard Business Review*, is a co-author of *No-Excuses Management* and *The Big Boys: Power and*

Position in American Business. He is a founding editor of *Fast Company,* a new business magazine based in Boston.

William Wiggenhorn is Motorola's corporate vice president for training and education and president of Motorola University. His contribution to this book received the 1990 McKinsey Award for best *Harvard Business Review* article.

INDEX